JOHN MASSON

THE

ATOMIC THEORY OF LUCRETIUS

CONTRASTED WITH MODERN DOCTRINES OF ATOMS AND EVOLUTION

Elibron Classics
www.elibron.com

Elibron Classics series.

ISBN 1-4021-4517-9 (paperback)
ISBN 1-4212-8868-0 (hardcover)

This Elibron Classics Replica Edition is an unabridged facsimile of the edition published in 1884 by George Bell and Sons, London.

THE ATOMIC THEORY OF LUCRETIUS.

THE ATOMIC THEORY OF LUCRETIUS

CONTRASTED WITH MODERN DOCTRINES OF ATOMS AND EVOLUTION.

BY

JOHN MASSON, M.A.

LONDON:
GEORGE BELL AND SONS,
YORK STREET, COVENT GARDEN.
1884.

The lense will not disprove
A present that eludes it . . . .
Though you saw the final atom-dance,
Making each molecule, that stands for sign
Of love, being present, where is still your love?

ROBERT BROWNING.

CHISWICK PRESS:—C. WHITTINGHAM AND CO., TOOKS COURT,
CHANCERY LANE.

PREFACE.

IT is strange that the Greek atomic theory, of which Lucretius is the sole exponent, has not, long before this, been set in a clear and detailed form before the English reader.

In Professor Veitch's little book ('Lucretius and the Atomic Theory,' 1875), only fifteen pages (pp. 10-25) deal with Lucretius's theory of atoms, and that only in a general way, while the rest of the volume is occupied with a very able criticism of modern Materialism. The scope of Professor Sellar's work does not allow him to enter at all minutely into the science of Lucretius, though his chapter on the connecting links between Lucretius's science and his poetry is most valuable.[1] Zeller has indeed given us in his 'Pre-Socratic Philosophy' an admirable sketch of the system of Democritus, but his account of the later development of the atomic theory in the hands of Epicurus is by no means equally complete. Lange's short chapters on Democritus, Epicurus, and Lucretius in his 'History of Materialism' contain acute enough criticism, though in his statement of facts Lange is by no means so trustworthy as Zeller. Neither Martha ('Le Poëme de Lucrèce,' 1873) nor Guyau ('La Morale d'Épicure,' 1881) attempt to give any complete or detailed account of the Epicurean theory

[1] We may also refer to the interesting chapter of Professor Sellar's 'Virgil,' tracing the influence of Lucretius's leading doctrines on the mind of the younger poet, and specially to the section on 'The Lucretian idea of Nature as it appears in the Georgics.'

of atoms, nor yet to point out its relations to modern science. In the present volume we have attempted to supply a short but careful account of the atomic theory as set forth by Lucretius,[1] and to show how far each of his propositions is in agreement with the conclusions of modern science, as represented by Clerk-Maxwell, Tyndall, and others. We have also endeavoured to point out the special vantage-ground of Epicurean science, and to show why it was possible for Epicurus's theory of the constitution of matter, as revived by Gassendi and others, to become the basis of modern physics, and to develop, stage by stage, into the atomic theory of modern chemistry.

To Lucretius the existence of atoms as an unchangeable basis of matter is necessarily connected with the fact of definite order and fixed laws in Nature. The crowning merit of Epicurean science was, as we have shown, that at so early a time it took so firm a hold of the principle of Law in Nature, —a fact grasped as firmly by Lucretius as it is by any modern man of science.

In modern scientific thought we find a parallel which helps us to realize how Lucretius's atomic theory taught him to regard Nature, and how his conception of Matter developed into a naïve theory of Evolution. Recent inquiry and speculation regarding the process of Evolution, the origin of Life and the potency of Matter, as illustrated by Tyndall's famous Presidential address, will enable us to realize more clearly, by comparison, what Lucretius's actual belief on these points was.

In explaining Lucretius's theory of the atomic structure of the soul, of the origin of consciousness, and of the method in which Will sets the body in motion, attention is called, so far

[1] See Dr. Brieger's review of our article in the 'British Quarterly,' Oct., 1875, on 'The Atomic Theory of Lucretius' ('Jahresbericht über die Fortschritte der class. Alterthumswiss'., 1877, 2nd part, pp. 63-5).

as we know for the first time, to the subtle part which atomic Declination plays in Epicurus's system. If it be thought that we have over-estimated the importance of this doctrine of Declination, which is stated and discussed at length in Chapter VII.,[1] we may quote the opinion of M. Guyau, who calls it, and we believe justly, 'the central and most original doctrine of Epicureanism.'[2] We have also pointed out the close relation between this doctrine of Lucretius and Professor Clifford's theory of 'Mind-Stuff.' The reasoning of both is based on the same principle, and both apply it with equal boldness. The question is an instructive one. In both cases, Materialism, finding itself hard pressed, escapes as it were by a back-door, and, in so doing, unconsciously confesses its own powerlessness to account, unaided, for the origin of Life and Thought.

M. Guyau devotes a long chapter of his very able work to the doctrine of atomic Declination, which he explains as implying a power of 'Spontaneity,' or modified Free-will action, residing in all forms of Matter, and by its working producing what we call Chance. He not only endeavours to prove that this was the actual teaching of Epicurus, but even accepts it as scientifically true. M. Guyau's theory of 'Spontaneity-in-Things' is one of the most remarkable castles in the air which the history of philosophy can show. We have examined it at length in an additional chapter.

We have thus attempted to give some account of Lucretius's position as regards both science and philosophy, and to indicate, as impartially as we can, both its strength and its weakness. It is hoped that the following pages may contribute a little

[1] In several papers we formerly attempted to indicate the philosophical consequences implied in Declination ('British Quarterly,' Oct., 1875; 'Journal of Philology,' vol. xii., 1883; 'British Quarterly,' April, 1882).

[2] 'Le point capital et vraîment original de la théorie épicurienne' ('La Morale d'Épicure,' 1881, p. 99).

towards a truer understanding of the Atomic Materialism of Lucretius, which forms none the less startling a chapter in the history of human thought, that we see it repeating itself, in somewhat subtler form, in the present day.

We must not forget to acknowledge, in common with all who have endeavoured to master the philosophic system of Lucretius, special indebtedness to Mr. Munro's edition. Much as Lachmann performed for the text, he left almost everything undone for the explanation of the poem,—a task of the utmost difficulty. It required qualities of mind which are rarely united to produce so trustworthy an edition as the great English one. While presenting our own rendering of the passages quoted, we have to acknowledge the constant aid derived from Mr. Munro's vigorous and admirably faithful translation. We are also indebted to Professor Fleeming Jenkin's thorough and original article on 'The Atomic Theory of Lucretius' ('North British Review,' vol. xlviii.), and have often quoted from it in the second and third chapters.

CONTENTS.

APPENDIX.

ERRATA.

P. 5. 'In 1873 a well-known chemist.' By an oversight Professor A. W. Williamson's name has been omitted.

P. 24. 'Two parts of hydrogen with one of oxygen,' ought to be—'Two parts *by volume* of hydrogen combine with one part *by volume* of oxygen' to form water. 'Parts' is now generally taken to signify parts by weight. Two parts by weight of hydrogen combine with sixteen parts by weight of oxygen to form water.

P. 69, note. 'The facts of observation quoted by Lucretius in support of this conclusion' as to the actual size of the sun. See the remarks in W. H. Mallock's 'Lucretius,' 1878, p. 38.

P. 119. In the quotation from Diog. L., 'The soul transmits sensation to the body which is in union with it,' &c., ought to be—'The body renders sensation possible for the soul which is born with it, and the soul, . . . by a faculty of its own, straightway realizes.' In note 1 on same page add—'Gassendi reads *κίνησιν* for *δύνησιν* in this passage, and translates the words *διὰ τῆς συντελεσθείσης περὶ αὐτῷ δυνάμεως*, per facultatem in ipso collaboratam.'—In first line of note read *ψυχὴ*, and in last line *οὖσα*.

P. 123. Text of ii. 251-93. At l. 257 *potestas* is the reading of Lachmann and Munro for *voluptas* of MSS. Lambinus transposes the last word of l. 258, and reads—

unde est haec, inquam, fatis avolsa voluntas,
per quam progredimur quo ducit quemque voluptas.

THE ATOMIC THEORY OF LUCRETIUS.

CHAPTER I.

THE ORIGIN OF THE ATOMIC THEORY.

OF late years the Roman poet Lucretius appears to have acquired a very strong interest for scientific men and others, and his name has found very frequent mention. This unwonted popularity is not on account of his bold attempt to abolish the gods and give a death-blow to superstition, hardly caring, meanwhile, whether religion might perish at the same time. Nor is he read by all even for his splendid poetic genius, for some of his admirers are extremely unpoetic people. The true reason is that his poem contains an admirably clear and straightforward exposition of a scientific theory which is now almost universally accepted, and which, in connection with Evolution, has gained a new and somewhat startling importance. The propositions in which Lucretius has stated his atomic theory anticipate some recent discoveries in both chemistry and physics in a most marvellous way. Science has now proved that his propositions as to the constitution of matter, in each case, are either certainly true, or else foreshadow the truth. Thus Lucretius's exposition of the ancient atomic theory has more than a mere historical interest. Indeed, its agreement with the results of modern science makes us wonder how the ancient students of nature, who had no means of verifying the observations of the senses through experiment, could have succeeded as they did. Like men walking abroad at night without a lantern, they could take with them no test of experimental inquiry by which to verify their hypotheses; but, in spite of all, some faculty enabled them to keep the right path. And this is the more wonderful, because (like our modern wave-theory of light and colour) the atomic hypothesis, in some points, goes altogether contrary to the evidence of the senses. Certainly, it must have been thought startlingly original when first proposed, nor is it easy to imagine what could have suggested to any man's mind a conception which the senses seem so to contradict. In these points it illustrates the fertile insight of the Greek mind. Yet while we accept this theory as to the constitution of matter as in very great part true, at the same time we reject as completely false Lucretius's deduction from it, the very thing for the sake of which he embraced it most eagerly. To Lucretius the existence of eternal uncreated atoms is important specially because this enables him to prove that the world has made itself, and that there is no room for Divine action within it.[1]

The poem on 'Nature,' *De Rerum Natura*, has an extraneous interest; it is of value for more than the thoughts of Lucretius. If Epicurus's great work, in thirty-seven books, entitled 'Concerning Nature,' or the other, 'Concerning the Atoms and

[1] The late Clerk-Maxwell, so famous as an inquirer in the domain of molecular physics, has even inferred from the character of the atoms, and the exact collocation of matter which they exhibit, the existence of a first cause, their Maker, from whom their powers are derived. Things which are unalterable cannot, he argues, have been formed by any of the processes which we call natural, and since each molecule is exactly similar to all others of the same kind, they bear the character of 'manufactured articles,' not of that which is eternal and self-existent. But, according to Professor Clifford, we have no evidence as yet that the molecules of any given gas are 'exactly' of the same weight. Moreover, even if they were, we have no evidence that it is absolutely impossible for molecules of matter to have been evolved out of ether by natural processes ('The First and the Last Catastrophe,' Essays and Remains, vol. i.).

Void,' still existed, in which he set forth his theory of atoms, we should go to him as the older and more original source. Not that even he was its author: the germ of the theory is attributed to Leucippus, whom Zeller considers to have been a contemporary of Anaxagoras and Empedocles. It was next taught by Democritus (sometimes called a pupil of Leucippus), who died about B.C. 360, and it was nearly a century later before it was fully developed by Epicurus. Zeller even says: 'It is certain that all the essential principles of the atomistic physics belong to Leucippus,'[1] and that Democritus derived them from him.[2] But Leucippus is only a name to us. Of Democritus we know far more. To whatever extent he may have accepted doctrines attributed to Leucippus, beyond question Democritus deserves the credit of having originated the atomic theory.[3] Democritus possessed a genuine scientific spirit. The facts of his personal history, his many journeys for the sake of acquiring knowledge and of observing nature, and his long life of laborious research, show that, as Zeller says, 'He was a man who with rare devotion gave his life to science, and who, as it is related, would have refused the kingdom of Persia in exchange for a single scientific discovery.'[4]

There can be no doubt that the whole framework of Epicurus's atomic theory and physics was built by Democritus, though in

[1] 'Pre-Socratic Philosophy,' vol. ii., p. 299 (Eng. Tr.).

[2] 'The fundamental conceptions of the atomistic physics, which are precisely those portions on which Lange lays so much stress, belong therefore to Leucippus, whom he passes over so unaccountably in silence.'—ZELLER, *ibid.*, vol. ii., p. 296 (Eng. Tr.).

[3] We may here refer to the acute and able chapter on Democritus in Lange's 'History of Materialism.' It however requires to be supplemented and corrected by Zeller's long section on Democritus, which is full of the most careful and impartial research.

[4] 'The writings of Democritus which Sextus [who lived in the third century, A.D.] still possessed were no longer in existence when Simplikius wrote.'—ZELLER, *ibid.*, p. 215. The fragments of Democritus which are preserved (Mullach, 'Fragmenta Philosophorum Græcorum,' vol. i.), are mostly from his ethical writings.

not a few points of importance Epicurus corrected and greatly expanded it. The atomic theory of Democritus was thoroughly unpopular among both the philosophers and the natural inquirers of Greece. Most of them pass it by contemptuously. Aristotle discusses it, but only to throw it aside. Therefore Epicurus deserves credit not merely for having developed and added to it, but also for the penetration which led him to revive, and thus to preserve for the world a theory which in his own day was generally scouted as absurd.

With the exception of a few letters and fragments of Epicurus, the works of all three are lost to us, and this most astonishing fruit of ancient thought, which has been adopted and substantiated by modern experimental science, is to be found fully described only in Lucretius's poem. Lucretius was born about the end of B.C. 99, and died in B.C. 55,[1] at the age of forty-four. His poem was written probably in the later years of his life. He has followed Epicurus closely, as coincidences with the fragments of Epicurus and the letters preserved by Diogenes, make very plain. He has added perhaps nothing really new to the theory: his contribution to it is only a most eloquent and distinct exposition of what he found in Epicurus. There is good reason for believing that but for this service done to it by Lucretius, Epicurus's system would never have exercised the powerful influence over modern thought which it has had. Judging from the careless, slipshod style and general formlessness of Epicurus's surviving writings, Epicurus could never have composed an account of his own doctrines, so clear and distinct, yet so concise as that which Lucretius has left us. In particular Lucretius's illustrations are admirable, so apt are they in each case to the point which he is explaining. In the eyes of science now, the value of Lucretius's poem lies in its full and exact statement of the ancient atomic theory as held by

[1] On the Ides of October, the same day on which Virgil attained the age of fifteen.

Epicurus. This it is which at present gives Lucretius so special an interest.

It was Gassendi who rescued Epicurus's atomic theory from the forgotten science of the old world and revived it as the truest basis for a scientific study of nature. Through Gassendi and his influence both on Newton and on Boyle,[1] as well as on many other minds in the 17th and 18th centuries, Epicurus's theory has taken firm root in modern science, and has developed, by stage after stage, into that atomic theory of modern chemistry which has proved fruitful in so many fresh discoveries made both in chemistry and, in our own day in particular, in molecular physics. The history of the Atomic theory in recent times is well known. The name of the chemist in whose hands it acquired a new force is now inseparably associated with it. Dalton assumed the existence of atoms, conjectured that the weight of the atoms making up each element is constant, assigned different specific weights to the different kinds of atoms, discovered the laws according to which they combine, and thus founded his celebrated Atomic Theory. So important were these discoveries and their results that Dalton has earned the title of the 'Father of Modern Chemistry.' The progress of chemical knowledge during the last century has been vitally connected with the hypothesis that there are such things as atoms, ultimate particles of matter, and its developments, nor is its value, as concerns fresh discovery, yet exhausted. In 1873 a well-known chemist, the President of the British Association, asked, in the course of his address, 'What is the meaning of the great activity shown at present in chemistry?' He answered the question thus: 'Chemists are examining the combining properties of atoms, and getting clearer views of the constitution of matter.' Professor Huxley says, 'If there is one thing clear about the progress of modern science, it is the tendency to reduce all scientific problems, except those that

[1] See the words of Boyle quoted by Lange, vol. i., p. 303.

are purely mathematical, to problems in molecular physics,—that is to say, to attractions, repulsions, motions and co-ordination of the ultimate particles of matter.' So important has proved Epicurus's conception that there exist such ultimate particles, molecules or atoms. Some of our readers may be surprised to find how similar the atom, as described by Lucretius, is to the modern chemical atom.

CHAPTER II.

THE ATOM OF LUCRETIUS.

BEFORE beginning to set forth his philosophy in due order, Lucretius expresses in the strongest way his obligations to his master: 'When human life lay shamefully grovelling upon earth, crushed down under the weight of *Religion, who showed her face from heaven, frowning upon mortals from on high with awful aspect,* a man of Greece was the first who ventured to lift mortal eyes to her face, and the first to withstand her openly.' Neither stories of the gods nor the thunders of heaven could make him afraid, but rather spurred him on, says the poet, to burst the bars of nature and find her secret. 'Therefore the living force of his soul prevailed, and he passed out far beyond the flaming walls of the world,[1] and traversed in mind the boundless universe, whence he returns, a conqueror, to tell us what can be and what cannot be; in short, on what principle each thing has its properties defined and its deep-set boundary-mark. *Wherefore religion is put beneath our feet and trampled on in turn; us his victory raises to heaven.*'

There is a boundless pity in the words describing the misery

[1] What would Lucretius have said to the spectrum analysis, by which the chemist can literally pass beyond the 'flaming walls of the world' (that is, the fiery circuit of ether forming our heavens), and bring us tidings from the distant stars? Wonderful, indeed, he would have thought it; but he would have valued it most if it could have aided him in any way to prove that the Gods have not created either the world or man, and are powerless whether for good or evil.

of men from the dominion of superstition,—the same pity and enthusiasm for humanity that has made saints and philanthropists in all ages, from Saint Francis to Robert Owen (though, certainly, there was more of the latter in the constitution of Lucretius). But we have quoted the passage to show what Epicurus was to Lucretius. Elsewhere he designates him a god; the popular deities, he says, are small compared with him. It is characteristic of the poet that, believing in no God whose help could avail mankind, he set up for worship the best thing that he could find,—a heroic man. But Lucretius is far more in earnest than he whom he delights to call his master. We cannot help questioning whether Epicurus would have approved of Lucretius's fervour even in the way of gratitude to himself. Was so great earnestness, even in the cause of his own philosophy, consistent with the calm and passionless tranquillity which the wise should seek?—This passage, moreover, gives the keynote to the whole poem. Probably nowhere does Lucretius forget that in his scientific inquiries the highest interests of man are vitally concerned. His demonstration of the existence of atoms is important chiefly, he feels, in order that man may be enabled to live his life aright, free from the crushing pressure of superstitious fears. It is science *for the sake of theology* that is here treated.

The first two books contain a number of propositions as to the qualities of the atoms, exactly what is denominated in our text-books the 'properties of matter.'

The first proposition is that 'nothing is ever begotten out of nothing by Divine power.' This outset is science and theology mingled, and it is, in this, characteristic of his whole work. 'Men see many phenomena take place in earth and heaven, the causes of which they cannot understand, and, therefore, believe them to be done by Divine power.' But I will show, says Lucretius, how all things are done 'without the hand of the gods.' Fervently, and with submission, as Lucretius realized the order of nature, the notion of deities *interfering* therewith must have seemed to him mean indeed. This, his first principle, holds true invariably of matter once created, as we daily observe it, and is assumed in every scientific treatise of to-day. By it Lucretius means to express that the laws of nature are constant, that phenomena take place according to well-defined laws, and that nothing happens without a cause for it in nature. His illustrations of the principle show that, at any rate, he had distinctly grasped the fact of law as few, or perhaps none, in his day can have done. If there is anything for which the world is indebted to Epicurus (probably still more to Lucretius), it is for a clear enunciation of the principle of law in Nature. Lucretius asserts it over and over again in modern-sounding phrases. '*It is absolutely decreed what each thing can do and what it cannot do, according to the conditions of nature.*'[1] Indeed, on this principle of the con-

[1] These words are often repeated in the poem and are intended to carry a very weighty meaning. Lucretius first uses them at the very outset of his task to express the triumphal and crowning result of Epicurus's intrepid researches into nature, viz., the knowledge of natural laws. It is this supreme discovery which, he tells us, finally delivers men from their bondage to superstition—

> unde refert nobis victor quid possit oriri,
> quid nequeat, finita potestas denique cuique
> quanam sit ratione atque alte terminus haerens.
> *quarė religio pedibus subiecta* vicissim
> *opteritur*, nos exaequat victoria coelo. i. 75-79.

We may compare the passage where he sums up the aim of his poem and, in order to do so, simply restates this one great principle, which he has throughout been tracing and proving to hold good through the whole of nature.

> cuius ego ingressus vestigia, dum rationes
> persequor ac doceo dictis, quo quaeque creata
> foedere sint, in eo quam sit durare necessum
> nec validas valeant aevi rescindere leges. v. 55-58.

At the close of the paragraph he recapitulates the same principle,

> quid queat esse,
> quid nequeat, &c. v. 88-90.

stancy of law his whole philosophy is based. As we shall see, Lucretius connects this principle of natural law with the indestructibility of the atoms:—were the atoms not unchangeable, the productions of nature would not obey definite unchanging laws. 'For, if the first beginnings of things could in any way be vanquished and changed, it would then be uncertain what could and what could not come into being, *in short on what principle each thing has its properties fixed, and its deep-set boundary-mark.*'[1] Lucretius grasps this principle of law in nature as strongly as does any modern man of science.

It need not be pointed out that this conception of the regularity and orderly sequence of natural phenomena is the first thing indispensable towards a scientific view of nature. But Lucretius makes the same mistake on this point as many modern scientific men,—that, if anything is said to be done by the hand of God, if, for example, He answers prayer, thereby 'a law is broken.' 'If, in consequence of prayer, external nature can be affected,' says the man of science, 'natural laws are thus at the mercy of man's volition, and no conclusion founded on their permanence is worthy of our confidence.'[2] So, to Lucretius, definite physical laws and the hand of God, acting in the world, seemed absolute contradictions. The possibility of connecting laws with a Divine power, with a God who 'thinks them progressively forth,'[3] seems never, even in the faintest, rudest shape, to have occurred to Lucretius. His only conception of Divine action is the polytheistic one of

[1] i. 592-7. Lucretius is constantly using the word *certus* to describe the *fixity* and *unchangeableness* of law, as manifested in the growth of natural productions. See the passages quoted in our chapter on Guyau's 'La Morale d'Épicure,' § 3.

[2] See Tyndall's essay on 'Prayer and Natural Law.' Tyndall's reasoning has been most searchingly examined by Dr. W. Ward. Dr. Ward's conception of a 'Divine pre-movement of events' is by no means a novel one, but he has worked it out with unusual force and grasp. We have tried to give some notion of his argument in the Appendix.

[3] Martineau.

interruption and interference. Lucretius's attitude of mind in regard to this forms a striking contrast to the instinctive God-consciousness of the Hebrew race, to whom the regularity of nature which we call law carried with it the inevitable inference of a Divine Will manifesting itself in this orderly fashion.[1] To Lucretius this regularity suggests just the opposite inference.

Lucretius's second proposition, which completes the first, is that 'nothing is ever annihilated, but all things on their dissolution go back into the first bodies,' that is to say, matter is imperishable, and the total quantity of matter is never diminished.[2] Lucretius closes his proof of the doctrine, after his wont, by a picture of the working of this principle in nature. 'The rains die when father Ether has tumbled them into the lap of mother Earth,' and in consequence the crops spring, the trees are covered with leaves and with fruit, men and animals are fed, the birds sing in the woods, the weak-limbed young of the herd gambol on the grass, 'intoxicated with the pure new milk,' and the children, human blossoms, make glad the city streets.

hinc laetas urbes pueris florere videmus.

A picture of all that is most fresh and cheerful in the world. The rains have passed away and out of sight, but they are not lost. 'Therefore,' Lucretius concludes, 'nothing that seems to be lost is utterly lost, since nature makes one thing afresh out of another, and suffers nothing to be begotten unless she has been recruited by the death of another.' The third pro-

[1] Thus, to quote one example out of many, in Psalm cxlviii., of the sun, moon, and stars, 'He hath made them fast for ever and ever; He hath given them a law which shall not be broken.' This thought is expressed in a very simple and natural way in Lord Houghton's beautiful little poem, entitled 'Good Night and Good Morning.' A child is watching the sun setting like a big red ball. She says 'Good Night' to the objects round her, and to the creatures all hastening to their homes, but not to the sun,

For she knew that he had God's time to keep
All over the world, and never could sleep.

[2] i. 215-264.

position states the existence of void. If there were no void, motion would be perfectly impossible in a world perfectly full of matter in its various forms, jammed together so that you could not insert a pin's point anywhere. Again, 'why do we see one thing exceed another in weight, though it be no larger in size?' Bodies of the same bulk ought to weigh the same, but do not. How can this be, unless the one has more empty pores in it than the other? Therefore, 'void exists mixed up with the substance of bodies.' Exactly this explanation is at present given of the different specific gravity of different substances. The third proposition states the existence of void, but for which motion would be impossible. The next two are, that all nature is made up of atoms and void, and that nothing else but matter and void exists.

The next proposition conducts us to the atom. 'Some bodies,' says the poet, 'are first-beginnings of things, the remaining bodies are formed from a union (*concilium*) of first-beginnings.'[1] These first-beginnings of things are the Lucretian atoms. He also very often calls them 'seeds' or 'seeds of things,' and 'bodies' or 'first bodies;' and sometimes 'first principles' (*elementa*), 'shapes' as they are conceived to differ from each other in form, and 'matter' as that from which things are made.[2] Anticipating a little, we may here try exactly to picture to ourselves an atom as Lucretius conceives

[1] Corpora sunt porro partim primordia rerum,
partim concilio quae constant principiorum. i. 483-4.

[2] According to Munro, Lucretius uses the following terms, *primordia* or *primordia rerum*, which is his 'proper and distinctive name for the atoms'; *semina* or *semina rerum* or *genitalia semina rerum*, a phrase not used by Epicurus, although Anaxagoras called his ultimate particles σπέρματα; *corpora* (answering to σώματα, which Epicurus appropriated to this special sense) or *corpora prima*; *elementa*; *figurae* (a phrase borrowed from Democritus), and *materies*. On these terms and their Greek equivalents, see Munro on i. 55. Zeller has pointed out that, amongst other names for his atoms, Democritus uses ἄτομα ('Pre-Socratic Philosophy,' vol. ii., p. 219, 3, Eng. Tr.).

it. It is a little hard kernel, perfectly solid and indestructible. 'The first-beginnings of things no force can quench ; they are sure to get the victory over it by their solid body.' Experience can give us no notion of such solidity. Everything we see around us in the world, however strong it may appear,—iron, stone, brass,—is yet destructible. Reason alone forces us to believe that the atoms are not. Ordinary bodies all have void within them ; but first bodies are perfectly solid. Without void 'nothing can be either crushed or broken up, or cut in two' (*nec findi in bina secando ;* Lucretius, who nowhere uses the word *atom*, by these words exactly translates the Greek ἄτομος). Without void, a thing cannot admit within it the destroyers, wet or cold or fire. Therefore the atoms, being impenetrable and indivisible, are indestructible. Lucretius is fond of calling them 'strong in their solid singleness,'

solida pollentia simplicitate.[1]

This is the most characteristic epithet which he gives them. Each atom is a distinct, separate individual. Matter cannot be divided farther, after you have reduced it to a collection of these individuals. Their 'singleness'—that meaning their distinctness of separate existence or individuality—is their strength. Though they enter into infinite fresh combinations, 'though stricken by countless blows through eternity,' they cannot be worn away. They are as perfect and fresh to-day as when the world was new. Each atom is perfectly hard, unchangeable and everlasting. (Of the more accessory properties of matter, it is proved that Lucretius assumes them to be elastic.)

As to the composition of this little kernel, though extremely small, it yet has parts ; each of these parts is 'of a least nature,' so small that it never has existed separate by itself, and will at

[1] i. 574. Again, at i. 612,

aeterna pollentia simplicitate.

no future time be able so to exist, since by its very nature it is a part of the other. These parts appear to be quite identical with one another. Each part is a *minimum*: nothing can be smaller than this and yet exist. These parts have existed from eternity side by side in the atom; 'in a close-wedged mass they fill up the composition of the first body.' 'The first-beginnings are not compounded from the union of those parts, but are to be considered strong in everlasting singleness.'[1] Lucretius appears to have thought three the smallest number of parts that an atom could have. Apparently he seems to have conceived each 'part' as representing an angle or corner,[2] so that an atom with three parts would appear to be a three-cornered or three-sided figure. The atoms differ in shape and size, and consequently in weight also, which must be in proportion to their bulk. As to shape, the atoms are not every one of them 'possessed of an equal size and like shape with one another.'[3] They differ widely in form.[4] Some are smaller. 'The subtle fire of lightning is formed of smaller shapes,' and can pass through openings better than 'this our fire, which is born of wood and sprung from pine.' Light is formed of smaller atoms than those of horn, and can therefore pass through it. Some atoms have hooks by which they are fastened together, and come closer to each other. Hard things, like diamond, basalt, iron, are formed of such atoms. 'Things which look to us

[1] primaque et una
inde aliae atque aliae similes ex ordine partes
agmine condenso naturam corporis explent, ...
sunt igitur solida primordia simplicitate
quae minimis stipata cohaerent partibus arte,
non ex illarum conventu conciliata,
sed magis aeterna pollentia simplicitate. i. 604-612.

[2] See Munro's note on l. 600. Epicurus wrote a book "On the Angle of the Atom'—περὶ τῆς ἐν τῇ ἀτόμῳ γωνίας.

[3] ii. 333-477.

[4] ἢ ἀγκιστροειδῆ, ἢ τριαινοειδῆ, ἢ κρικοειδῆ. Plutarch, 'De plac. ph.,' i., 3. Is there any other authority for 'ring-shaped' atoms?

hard and close-textured must consist of atoms that are more hooked together and must be held in union, because welded together through and through out of atoms that are, as it were, many-branched. Amid this class in the foremost line stand diamond-stones, accustomed to despise blows, and stout basalt blocks,[1] and the strength of hard iron, and the brass bolts which scream out as they hold fast in the doors.'[2] Liquids, are, as a rule, formed of smooth and round elements, but a sluggish fluid, like oil, may have its atoms 'larger or more hooked and intertangled' than those of wine. In general, things which gratify the senses are formed of smooth and round atoms; whatever is painful and harsh, its elements are more hooked and rough. Again, 'some elements are with justice thought to be neither smooth nor altogether hooked with curved points, but rather to have very small angles slightly projecting, so that they can rather tickle than hurt the senses,' for example, tartar of wine and elecampane. Apparently Lucretius supposes the different shapes of the atoms to result altogether from the position in which the least parts are placed within each. 'Every different arrangement of the parts yields a different manner of form of the atom.' But there is a limit to these differences: the number of shapes is finite,[3] but the atoms of each shape are infinite in number. Lucretius argues

[1] '*Silices* denotes the hard blocks of volcanic basalt with which the Romans paved their streets and roads.'—MUNRO.

[2] Denique quae nobis durata ac spissa videntur,
haec magis hamatis inter sese esse necessest
et quasi ramosis alte compacta teneri.
in quo iam genere in primis adamantina saxa
prima acie constant, ictus contemnere sueta,
et validi silices ac duri robora ferri
aeraque quae claustris restantia vociferantur. ii. 444-50.

[3] ii. 478-494. In stating this, Lucretius supposes an atom formed of three least parts, and adds that 'you may increase them by a few more.' These words may be meant literally or not. It is calculated that from three parts 6 different shapes might be derived, from four 24, from five 120, from six 720, from seven 5,040.

that there is not an infinite number of differently shaped atoms, for if so, some atoms must be infinitely large ; moreover, some new kind of thing would be constantly coming into existence, something more brilliant, more beautiful, more fragrant, more melodious, or the opposite, than any before-existing thing.[1] Epicurus held that the number of different shapes, though not infinite, was inconceivably great. Lucretius merely proves that it must be finite. All atoms of the same shape are not necessarily of the same size. There are both greater and smaller among atoms of the same shape.[2] Again, some round atoms are smoother in surface than others also round.[3]

As to size, we must keep well in mind that the atom, as Lucretius conceived it, is a very tiny body. 'The whole nature of the first-beginnings,' he says, 'lies far beneath the ken of sense.' Early in the First Book he proves, by illustrations to which we shall afterwards refer, that 'Nature works by bodies which are invisible.' This is why he so often uses the epithet 'blind,' that is invisible, of the atoms and their movements. But he insists emphatically that the atoms are not infinitely small.

Lucretius did to a certain extent try to realize the size of his atoms. From his theory of sight, we see that he believed the atoms to be exceedingly small, as compared with the smallest visible object. According to Epicurus, sight is caused by a succession of images of any object striking on the eye.

[1] ii. 478-521.

[2] Thus Lucretius tells us that 'the heavenly fire of lightning, subtle as it is, is formed of smaller shapes, and therefore passes through openings which this fire of ours cannot pass through, being born of logs and sprung from pine' (ii. 384-7)—these lines meaning that the atoms of fire are of the same shape (that is, spherical) as those of lightning, but larger; cf. vi. 225-7. The older atomists certainly held that atoms of the same shape differ in bulk. Democritus held that fire is composed of small round atoms. Theophrastus ('De Sensu,' c. 75) tells us that Democritus explained red as produced by the juxtaposition of atoms similar to those of fire, but larger,—ἐρυθρὸν δ' ἐξ οἵωνπερ τὸ θερμὸν, πλὴν ἐκ μειζόνων.

[3] ii. 469.

These 'images,' which are films so thin as to be utterly intangible, are made of atoms, and are constantly streaming from the surface of every object. 'In a moment of time there must be carried away from the surface of objects, images many in number, in many ways, in all directions round.'[1] These films, intensely thin,[2] never cease for a single moment to fly away from every object, and the air is ever full of them, but they are apparent only to the eye, and the substances which send them off show no trace of any loss; how small, then, must be the atoms forming such 'images.' Again, 'the mind perceives much thinner images' than the eyes can.[3] These images, which are seen by the mind's eye, whether in thought or in sleep, are infinitely finer even than those perceived by the eye, yet these mental images, too, are atomic. How infinitely fine, then, must be the atoms forming them! Again, says Lucretius, think of any very small living creature, and then of its entrails or heart or eye. Exceedingly small such an organ must be, yet it is formed of a number of atoms. Further, the animal life or soul of such an animalcule (according to Epicurean psychology the soul of any creature is vastly finer and lighter than its body), how almost inconceivably small, Lucretius reminds us, must the soul of such a creature be in proportion to its tiny body, yet its soul must be composed of many atoms.[4]

[1] sic ab rebus item simili ratione necessest
temporis in puncto rerum simulacra ferantur
multa modis multis in cunctas undique partis. iv. 163-5.

[2] suptili praedita filo. iv. 88.

[3] scire licet mentem simili ratione moveri,
per simulacra leonem et cetera quae videt aeque
nec minus atque oculi, nisi quod mage tenuia cernit. iv. 754-6.

[4] 'Come, now, and learn how thin is the nature of which an image is formed. First of all, since the first-beginnings are so far below the ken of our senses, and so much less than those which our eyes first begin to be unable to see, learn in a few words how fine the germs of all things are. First, living things are, in some cases, so very small that their third part cannot be seen at all. Of what size are we to think the gut of these

Lucretius implies that the atoms may be even smaller than one would infer from such considerations. Lucretius's psychology, and, indeed, his whole system, of necessity demands that his atoms should be exceedingly minute.

corporibus coecis igitur natura gerit res.

Sir William Thomson says that if a drop of water could be magnified to the size of our globe, the molecules comprising it would appear to be of a size varying from that of shot to that of billiard-balls. According to Clerk-Maxwell, about two million molecules of hydrogen placed in a row would occupy 39-1000th of an inch, and a million million million million of them would weigh something more or less than 70 grains troy. We question whether Lucretius would have assented to his atoms being rated at so small a size as this. In conclusion, Lucretius denies to the atoms all secondary qualities, which he sharply distinguishes from essential properties.[1] They are colourless. They are not white or black or azure because existing things are white or black or azure. All colours can change into other colours, but that which changes is perishable, therefore the atoms are not endowed with colour. It *is* possible for us to conceive of atoms colourless, just as 'men who are born blind can yet recognize bodies by touch, though from the first they have never been associated in their minds with colour.' Neither have they sound, or scent, or warmth, or cold. All such qualities belong to things which are perishable; but 'they must all be withdrawn from the first-beginnings, if we wish to assign for existing things imperishable foundations, for the safety of the universe to rest upon, that you may not have things returning altogether to nothing.' Lastly, the atoms are void of sense—

creatures to be, or their heart, or eyes, or their limbs, or any part of their frame? How small are these! What, then, must be the size of the several first-beginnings, whence their soul and the nature of their mind is formed? See you not how fine they are, how minute!' iv. 110-22.

[1] ii. 730-864.

mere dead matter.[1] Thus all their characteristics are here summed up. This, then, is the Lucretian atom, tiny yet so strong; after it has taken part in innumerable combinations, which have been formed and broken up and formed again, it remains fresh and perfect as ever.

It is interesting to know what was the reasoning by which Lucretius arrived at the result of ultimate atoms and their properties. He gives nine or ten arguments[2] to prove either that there are atoms 'of solid singleness,' or that the atoms are indestructible: merely two forms of expressing the same statement. His reasoning is somewhat as follows:—

In the first place, Lucretius holds that, admitting the existence of matter and void, each of these must of necessity exist 'by itself and unmixed.' For, wherever void is, there matter cannot be; and wherever body is, there void cannot be. That is to say, from the existence of void, absolutely empty space, Lucretius infers the existence of its opposite, the not-void, perfectly solid matter. Again, things, it is admitted, have all void within them; but how could they hold it in and continue to keep it within them, unless their substance was perfectly solid, pure, unmixed matter? Thirdly, having no void within them, the atoms must be indestructible. (Here Lucretius incidentally throws out a proposition as to the destructibility of bodies. The reason why bodies of any kind are destructible is, that they are not perfectly solid. 'The more of void anything contains within it, the more completely does it give way before any assault.') It is here—in its perfect solidity—that Lucretius's atom differs most from that of modern chemists, who, as Professor Clifford says, explain the hardness of solid matter 'by the very rapid motion of something which is infinitely soft and yielding.' Lucretius has no notion of this. He argues next that, admitting solid atoms, you can explain the existence of soft bodies, such as air, water, earth, by the

[1] ii. 865-990. [2] i. 498-634.

admixture of void; but if your atoms are destructible and soft, how can the existence of hard bodies, like basalt and iron, be explained? On the other hand, assuming the atoms to be solid, their closer union will account for the utmost strength exhibited by any substance.[1] Moreover, Lucretius is persuaded that, as Professor Jenkin puts it, there is an immense 'wear and tear going on' in Nature; if the atoms were at all frail, 'it is not consistent that they could have continued from eternity, though stricken and tossed about eternally by countless blows.' To sustain these fearful shocks, the strain of eternal combinations from atoms to things, and dissolutions from things back to atoms,—'under that strong pressure within the very jaws of death' Lucretius says[2]—there must be indestructible first-beginnings.

The sixth reason is an important one. We give it at more length, and in the poet's own words: 'Had nature set no limit to the breaking of things, the bodies of matter would by this time have been reduced so far by the breaking of past time, that nothing could be conceived out of them and reach its full growth within a fixed time.' 'But now, without doubt, a limit has been set to their breaking, and abides sure, *since we see each thing produced afresh, and, at the same time, well-defined periods fixed for things, each after its kind, to reach the flower of their age.*' That is to say, we see in all the productions of Nature that matter obeys definite unchanging laws; therefore, in order to produce these regular results, the ultimate basis of matter must be definite and unchangeable. Thus Lucretius deduces this property of the atoms from his great principle of law in nature, as illustrated by the regular periods within

[1] sunt igitur solida pollentia simplicitate,
quorum condenso magis omnia conciliatu
artari possunt validasque ostendere viris. i. 574-6.

[2] nam quid in oppressu valido durabit eorum,
ut mortem effugiat, leti sub dentibus ipsis. i. 851-2.

which growth and life go on. Lucretius justly feels the last to be a strong argument, and he repeats it in a slightly varied form: 'Since there abides a limit of growing and retaining life, assigned to things, each after its kind, and since by the laws of nature it stands inviolably decreed what they each can do and cannot do, and since nothing is changed, but all things are so constant that the different kinds of birds, all without intermission, exhibit on their body the distinctive marks of their species, they must, without doubt, also have their bodies formed of unchangeable matter. *For if the first-beginnings of things could in any way be vanquished and changed, it would then be uncertain what could and what could not spring into being; in short, on what principle each thing has its properties fixed, and its deep-set boundary mark; nor could the generations so often reproduce, each after its kind, the nature, habits, way of life, and motions of the parents.*'[1] Thus he again deduces the properties of the invisible atoms from the character of existing things which we can see,—for do not these represent the powers of the atoms? From the constancy of all the phenomena of Nature (as illustrated by the distinctive marks, habits, and motions of

[1] Denique iam quoniam generatim reddita finis
crescendi rebus constat vitamque tenendi,
et quid quaeque queant per foedera naturai,
quid porro nequeant, sancitum quandoquidem extat,
nec commutatur quicquam, quin omnia constant
usque adeo, variae volucres ut in ordine cunctae
ostendant maculas generalis corpore inesse,
inmutabili' materiae quoque corpus habere
debent nimirum. nam si primordia rerum
commutari aliqua possint ratione revicta,
incertum quoque iam constet quid possit oriri,
quid nequeat, finita potestas denique cuique
quanam sit ratione atque alte terminus haerens,
nec totiens possint generatim saecla referre
naturam, mores, victum, motusque parentum. i. 584-98.

The terms *reddita constat*, *sancitum extat*, are intended specially to denote the fixity of natural law.

the various species of living things), he infers that the atoms are unchangeable. Lastly, if Nature allowed of division beyond the atom, if matter were infinitely divisible, then nothing could be reproduced out of such least parts, because particles which are infinitely small 'cannot have the properties which birth-giving matter ought to have, that is to say, the various entanglements, weights, blows, clashings, and motions, by means of which things severally go on.'[1] Exactly to the same effect Clerk-Maxwell says: 'We do not assert that there is an absolute limit to the divisibility of matter: what we assert is, that after we have divided a body into a certain finite number of constituent parts called molecules, then any further division of these molecules will deprive them of the properties which give rise to the phenomena observed in the substance.'[2] Thus Lucretius reasons with perfect consistency from the eternal youth of nature, the freshness and perfection of all natural objects, generation after generation and growth after growth, to the indestructibility of the atoms forming them. He sees in the 'solid singleness' of the atoms an adamantine wall against which death and decay recoil—an eternal barrier-line which reason discerns looming far, far beyond the phenomena of the changing world. This alone, he believes, saves nature from degeneration.

It has repeatedly been said that Epicurus's hypothesis of the atoms was a 'mere guess.' We think that hardly any one, who has read attentively the above abstract of Lucretius's argument, will agree to the statement that his Atomic theory is but 'a guess.'

This theory of Lucretius that there really are such things as atoms, ultimate indivisible particles of matter, is now accepted. The modern chemist believes also, like Lucretius, in a limited number of different atoms, from each of which he

[1] i. 628-34.

[2] 'Theory of Heat,' p. 285, quoted by Munro.

supposes an elementary chemical substance to be composed. It is indeed strange to think what could have first suggested to any man's mind a theory so different from what the senses tell us. It has been thought by some a mere conjecture, suggested by the sight of the sunbeam kindling the countless motes afloat in the air. Lucretius's arguments prove at once that this cannot be. A single glance at the principal reasons by which modern science has arrived at the atom, enables us to see how just Lucretius's reasoning was. We may give the general principle in Newton's own words: 'All things considered, it seems probable that God, in the beginning, formed matter in solid, massy, hard, impenetrable particles, of such sizes, figures, and with such other properties, and in such proportions to space, as most conduced to the end for which He formed them.' 'While the same particles continue entire, they may compose bodies of one and the same texture in all ages; but should they wear away or break in pieces, the nature of things depending on them would be changed. Water and earth composed of old, worn-out particles would not be of the same nature and texture now with water and earth composed of entire particles in the beginning. And, therefore, that nature may be lasting, the changes of corporeal things are to be placed only in various separations, and new associations and motions of these permanent particles.' Balfour Stewart uses language that might have come from Lucretius himself when he says, 'A simple elementary atom is a truly immortal being, and enjoys the privilege of remaining unaltered and essentially unaffected by the powerful blows that can be dealt against it.'[1] Thus we are justified in conceiving of the atoms as 'the foundation-stones of the universe,'[2] which amid all the changes of matter ever remain 'unbroken and unworn.'

Therefore the song of nature over her task is,—

[1] 'Conservation of Energy,' p. 7.

[2] Clerk-Maxwell.

'No ray is gone, no atom worn,
My oldest force is good as new,
And the fresh rose on yonder thorn
Gives back the bending heavens in dew.'

If it were not so, were there not indestructible atoms, the rose that opened its dewy leaves to the sun this morning could not be as fresh and pure and fragrant as the first rose that ever opened its petals on this earth ; sunlight and air could not be as bright and fresh, the human form as fair, all the world as beautiful, life as keen, and the longing in the heart of the youth to enjoy existence to the full as strong as it was a thousand years ago. Lucretius saw as clearly as Newton did that, while the atoms hold fresh and unalterably 'strong in their everlasting singleness,' though the bodies they compose should waste away, still, completely fresh and new ones, as strong and as perfect, may be formed when they unite again.

The laws of chemical combination furnish a most powerful argument, and clearly prove that the atoms are unalterable. The chemist finds, for example, that oxygen produced from any source, from air, water, or rocks of any geological period, and hydrogen produced from any source, such as water, coal, or meteoric iron, combine together, always in exactly the same proportions, namely two parts of hydrogen with one of oxygen, to form water. He asks then confidently, 'if during the whole previous history of either specimen, whether imprisoned in the rocks, flowing in the sea, or careering through unknown regions with the meteorites, any modification of the molecules had taken place,' is it possible that these two gases could combine always in exactly the same proportions ?[1] Decidedly they could not. The spectroscope again proves that molecules of any element, such as hydrogen, always give exactly the same set of periods of vibration, but if the little particles which vibrate at exactly this speed could be broken or altered in any way,

[1] See Clerk-Maxwell's Lecture on 'Molecules.'

they could no longer vibrate at this certain fixed speed and thus produce the corresponding line in the spectrum which is the sign of hydrogen. (Lucretius, by the way, would not have allowed the possibility of vibration, implying internal movement, in his atoms.) Both the last examples are only special applications of the same general principle which Lucretius realized so clearly, and would, as such, have been welcomed warmly by him. That principle is, that 'if matter really obeys definite, unchangeable laws, the ultimate materials employed to make matter must themselves be definite and unchangeable.'[1]

Some of Lucretius's assertions regarding the atom may appear rather arbitrary,—for example, the statement that atoms have parts. His conception apparently is, that anything which has extension and exists in space must have parts. Intellectually we can conceive these parts, though really the atom is indivisible. 'Since these parts cannot exist by themselves, they must needs cleave to that from which they can in no wise be torn away. First-beginnings, therefore, are of solid singleness, massed together and cohering closely by means of least parts, not compounded out of a union of these parts, but, rather, strong in their everlasting singleness. From the atoms nature no more allows anything either to be torn loose or to

[1] Clifford uses the following illustration in proof that the doctrine of atoms must no longer be called a theory, but is a simple statement of the facts. Imagine, he says, a room full of fiddles hanging up with strings tuned to vibrate to certain notes. Sing one of those notes, all the fiddles answer. Compress the fiddles so that they are in contact, they will not answer to the notes. Bring them closer, but still so as not to be in contact, they will still answer exactly to the note as before.—So, if we compress a gas within certain limits, the rate of vibration which belongs to the gas is still not altered. This shows that the rate of vibration does not belong to it as a whole, but belongs to the individual parts of it. 'By such reasoning as this,' he continues, 'it seems to me that *the modern theory of the constitution of matter, is put upon a basis which is absolutely independent of hypothesis. The theory is simply an organized statement of the facts.*'

be worn away, reserving them as seeds for things.'[1] Thus, the individuality of each several atom, as Lucretius conceives it, is like human individuality in its distinctness. Perhaps by describing the parts of the atom as cohering closely without any intervals of void, he meant his reader to realize strongly how different is the atom in structure from matter as we have experience of it. But he offers no proof of this dogma, and his acceptance of it from Epicurus simply shows how closely he followed his master in the domain of physics. On the remarkable hypothesis of atoms with 'hooks,' Newton comments thus, 'The parts of all homogeneous hard bodies which fully touch one another, stick together very strongly. And for explaining how this may be, some have invented hooked atoms, which is begging the question.'[2] It is certainly difficult to conceive of atoms, encumbered with such an apparatus as hooks and eyes, being indestructible. Perhaps it is easiest to conceive of round atoms being indestructible. A recent writer, Professor Veitch, criticises the notion of hooked atoms to the same effect as Newton. 'These conceptions,' he says, 'are obviously simply and rudely mechanical. They may be taken as, at the utmost, very dim foreshadowings of gravity and cohesion. The ideas of polar attraction and repulsion and of chemical affinity have no place.'[3] The con-

[1] quae quoniam per se nequeunt constare, necessest
haerere unde queant nulla ratione revelli.
sunt igitur solida primordia simplicitate
quae minimis stipata cohaerent partibus arte,
non ex illarum conventu conciliata,
sed magis aeterna pollentia simplicitate,
unde neque avelli quicquam neque deminui iam
concedit natura, reservans semina rebus. i. 607-14.

'*ex ordine*, having each so existed without possible shifting of position.'—Munro. On the parts of the atom, see also ii. 483-94, and our note on p. 15.

[2] Quoted by Munro on ii. 455.

[3] 'Lucretius and the Atomic Theory,' p. 39.

ception of atoms with hooks attaching them to each other is, no doubt, rude enough, as Professor Veitch remarks, but as a metaphor, it is not beneath the use of modern science, as the following passage shows:—'Suppose we were to begin teaching chemistry by saying that carbon is made up of atoms which have four hooks or hands by which they can hold on to other atoms; that oxygen atoms have two hooks, and hydrogen atoms one. Consequently we can hook two hydrogen atoms to an oxygen atom, and this makes water; or we can hook two oxygen atoms to a carbon atom, making carbonic acid; or we can hook four hydrogen atoms to a carbon atom, making marsh gas. . . . These statements belong to the doctrine of atomicities. Nobody doubts that the statements represent, in highly metaphorical language, real facts of chemical action.'[1] Further, we may ask in passing, is it correct to say that Lucretius has no notion of chemical affinity? If our interpretation of the term *concilium* be correct,[2] he had, at all events, some faint one.

Having arrived at the atom, Lucretius proceeds to discuss rival theories as to what is the original element of all things. The Epicurean criticism of other systems was by no means of the fairest nor yet of the best-informed, as we see when Lucretius utterly misrepresents the 'fire' which Heraclitus assumes as the sole element of things, by discussing it as if it were ordinary fire. On no other system is Lucretius so severe as on that of Heraclitus, whose popularity he scornfully explains by the fact that 'fools like dark sayings.'[3] The reason of so great a passion against this philosopher is not hard to find. The physical system of the Stoics, the enemies of Epicurus, was derived from Heraclitus. Lucretius professes

[1] Professor Clifford's 'Essays and Remains,' vol. ii., p. 307.

[2] See Chap. III.

[3] In this opinion of Heraclitus, Lucretius does not agree with the estimate of Professor Ferrier—'the deepest probably, if also the darkest, of all the thinkers of antiquity' ('Lectures on Greek Philosophy,' § 4).

to see very clearly where all systems that conflict with his own are defective. Their founders 'have gone to wreck on the first-beginnings of things, and there, though great, they have fallen with great downfall.' Space will allow us to refer only to his elaborate and vigorous refutation of Anaxagoras. One doctrine of Anaxagoras, adopted by the Peripatetics, was in direct hostility to his own theory. In combating it, Lucretius defends the Epicurean side in a controversy of the day. Anaxagoras and the Peripatetics held that the parts of a body are in every respect similar to the whole ; that flesh is formed of minute fleshes, blood of minute drops of blood, earth of minute earths, gold, water, of minute particles of gold and water. This doctrine (with other similar ones) was called in later times Homoiomereia, that is, the 'likeness of parts (to the whole).' It is not hard to see why the atomic theory is at enmity with this. If the one holds, the other must utterly fall to the ground. Suppose you take a grain of earth and divide it again and again. So long as the parts are visible, they possess properties similar to the whole grain. They are still recognizable as earth. Even after the parts are so small as to be no longer visible, we can still conceive of the process being carried on by some finer instrument. The question then occurs, Can this subdivision be repeated for ever? The atomists answer, It cannot. After it has been divided a certain number of times you will come to parts extremely small, which are impenetrable, no longer divisible, things which cannot be cut, that is to say, Atoms. According to Anaxagoras, on the other hand, this process may be repeated for ever. Every smallest subdivision of the grain of earth is still like the whole grain, and you may repeat the process of division without ever coming to an end. Thus the two doctrines were in direct hostility.

Lucretius now proves at length that matter and space must both be infinite. If there be limits to the universe, then all matter must long ago have collected in a heap on the floor of the limited space. So, too, he will not hear of a centre of the universe towards which all things tend. 'There can be no centre [where the universe is] infinite,'[1] and if there were one, he continues, there is no reason why things should be attracted any more than repelled by it. In this point Epicurus stands in marked opposition to Aristotle, whose whole conception of the universe is based upon the notion that it has a centre which is the goal of all motion. Matter, too, must be infinite. If space were infinite and matter finite, in the first place the atoms could never have been able to combine so as to produce anything, but would have been carried apart and scattered throughout the infinite void. Thus the atoms fit to produce the world could never have come together. Another reason introduces one of the most characteristic conceptions of Epicurean science. According to Epicurus, there is a continual waste going on in the world. It is an organism which is continually losing part of its substance by natural waste, and moreover the 'blows' of the atoms which keep constantly beating on every substance tend to weaken it and break it up. Further, 'the air lying round about always beats on things'[2] and helps to wear them out. 'All things perish when they have been rarefied by the ebb of particles, and succumb to blows from without,' since 'the atoms never cease to destroy anything by thumping it from without, and to attack and overpower it by blows.'[3] In consequence of this loss of

[1] i. 1070. [2] vi. 1028. Cf. iv. 932-4.

[3] ii. 1143-7. Lucretius speaks at times of these 'blows' as helping to hold the world in existence, as at ii. 528-31,

probavi
versibus ostendens corpuscula materiai
ex infinito summam rerum usque tenere
undique protelo plagarum continuato.

'I have shown that the minute bodies of matter do continually uphold the world through an unbroken succession of blows on all sides.' The word *protelo* 'appears to denote a number of draught-oxen, yoked one substance, and of the inroads made by the strain of attacking forces, the world requires to be fed by a fresh stream of atoms, constantly flowing in from the infinite void. 'As the constitution of living creatures, when deprived of food, loses substance and wastes away, even so the world must be dissolved as soon as matter has ceased to be supplied, being in any way diverted from its course.'[1] The mere impact of the ever-moving atoms, beating on the world from without like the shock of waves on the side of a ship, is not simply and merely destructive. To a certain extent Lucretius thinks it may even help to hold things together, for a time,[2] but if this great world is to maintain its being and action, it must be constantly fed with matter from the infinite sum without. A precarious condition, for how easily might the supply of matter 'somehow lose its way' in the immensity of the void!

materies aliqua ratione aversa viai.

(As we shall see, the direction of these atoms coming to feed the world is *upwards*.[3]) In consequence of collision the atoms which naturally fall downwards are forced up, but there can be no constant succession of upward-streaming atoms unless

in front of the other, and advancing by even, successive pulls; hence it well expresses the effect produced by the continuous succession of blows of atoms' (Munro). In such a passage as this 'blows' stands for its result, viz., the supply of fresh matter, which, without these atomic collisions, could not rise upwards to feed the world. The reasoning of i. 1041-51 is conclusive as to this.

[1] nam veluti privata cibo natura animantum
diffluit amittens corpus, sic omnia debent
dissolvi simul ac defecit suppeditare
materies, aliqua ratione aversa viai. i. 1038-41.

[2] 'Blows from without cannot hold together all the sum' [*i.e.* this world], though 'they can frequently strike upon and detain a part, until others come, and the full sum of matter can be completed.' i. 1042-5. (*Plagae* occurs in two different meanings at ll. 1042 and 1045.) On this subject, see ii. 1105-50.

[3] See page 48, and note.

matter be infinite. The medium through which this loss and gain come to things is the air; 'Whatever ebbs from things is all borne into the great sea of air, and it in return gives back particles to things.'[1]

This notion of a loss of Energy constantly going on from the world has much in common with the scientific doctrine of the Dissipation of Energy. Thus Professor Balfour Stewart shows that each form of energy is not capable of being transformed directly, so far as we know at present, into every other form. Thus, for instance, energy of visible motion cannot be *directly* transformed into energy of chemical separation or into radiant energy. The fact that heat can only be transformed into mechanical energy subject to a certain condition, shows us that there must be constantly going on a Dissipation of Energy. So 'if we could view the Universe as a candle not lit, then it is, perhaps, conceivable to regard it as having been always in existence; but if we regard it rather as a candle that has been lit, we become absolutely certain that it cannot have been burning from eternity, and that a time will come when it will cease to burn.'[2] Thus Lucretius's belief that old age and death must necessarily come to the world when its waste becomes greater than its supply, is not without scientific basis.

What then would be the consequence supposing the influx of atoms from the infinite were to stop? Lucretius states it thus,—'Swift as flame the walls of the world would suddenly break up and fly asunder along the mighty void, and for the same reason all other things would follow: all the heaven from its inmost quarters would tumble down, and in an instant the earth slide from beneath our feet and wholly pass away along the boundless void, the ruins of the heaven and of earthly things all wildly mixed, and the atoms unloosed from their bonds of union, so that in a moment not a wrack shall

[1] v. 275-8. [2] 'Conservation of Energy,' chap. v.

be left behind, nought save lone space and the unseen first-beginnings. For on whatever side atoms shall first be wanting, this side will be the gate of death for things in being.'

It is curious how often Lucretius refers to the destruction of the world.[1] The earth, he says, is ceasing to bear with its former fertility: it is manifestly grown old. Even his own generation, he thinks, may see the end of all things.[2] No doubt the frequency of severe earthquakes and volcanic eruptions in Italy and Sicily may have impressed him vividly, and led to such anticipations.[3] But there were far more cogent reasons in his own theory of Nature. Lucretius seems to have been keenly conscious of the truth that a chance-made world could not well be permanent, that the chance-moved atoms might fly from their orderly combination and undo the world even more readily than they had joined to upbuild it. Considering that he held such a theory of the world's origin, must

[1] See i. 1102-13; ii. 1148-74; v. 91-109, 338-46, and 364-75. In the last passage, the lines,

> neque autem corpora desunt,
> ex infinito quae possint forte coorta,
> corruere hanc rerum violento turbine summam
> aut aliam quamvis cladem importare pericli

remind us of the remarkable theory of 'Spontaneity-in-things,' attributed to Epicurus by M. Guyau. Such an insurrection of atoms out of the infinite *might* be conceived thus to arise and imperil the world in consequence of the power of declination, since atoms in the void are not subject to the laws which restrict and nullify atomic declination in masses of matter. In a storm of wandering atoms, 'Spontaneity' would be a mighty and dangerous force.

[2]
> sed tamen effabor: dictis dabit ipsa fidem res
> forsitan et graviter terrarum motibus ortis
> omnia conquassari in parvo tempore cernes. v. 104-6.

[3] 'They told us, if I remember, that they had an earthquake on this part of the coast of Italy about once every five years. Italy is a land of volcanoes, more or less subdued. It is a great grapery, built over a flue.' —'Lord Byron and his Contemporaries,' by Leigh Hunt, chap. iv.

he not indeed have been surprised that the torrent of atoms had not burst loose many a year before. For, in a world of purely atomic origin, must not any rent or flaw grow ever wider and deeper, with accelerating increase? for where is there any recuperative power, once the elaborate atomic combination has met with any jar or crack? Such a world resembles in its structure a great many-sided crystal which, so soon as the slightest flaw has touched it, at once falls into fragments and dissolves. If a world like this had yet held together for so long, must not its destruction have seemed to Lucretius now inevitably nigh at hand?

Moreover, Lucretius conceives the world as liable to destruction because it exists only in virtue of a certain balance[1] or equipoise of natural forces. In part the atomic motion tends to preserve things in being, and in part it tends to break them up. These opposing principles show themselves visibly every day around us in birth and death, in growth and decay, but they are at work everywhere. Thus mightiest forces, attacking it within and without, are ever seeking to destroy the world,—forces as mighty are ever at work combating these and restoring their inroads. So long as this balance of warring powers holds, so long at least as the destructive forces do not gain the upper hand, so long the world will last.

> sic aequo geritur certamine principiorum
> ex infinito contractum tempore bellum.

But this warfare may at any time come to a close.

The First Book concludes with these words,—to the disciple who will earnestly ponder his teaching 'one thing shall grow clear after another, nor shall the blind night rob thee of the road that thou see not to the full the most secret ways of nature: so truly will one thing light the torch for another.'

[1] The principle illustrated by this equilibrium of antagonistic forces is an important Epicurean doctrine, under the name of ἰσονομία, or 'balance.' See Book v. 381-96; ii. 569-80; Cicero, 'De Natura Deorum,' i. 50.

CHAPTER III.

THE ATOM OF LUCRETIUS (CONTINUED).

THE Second Book begins with the well-known lines, *Suave, mari magno,*—

> '’Tis pleasant, when the seas are rough, to stand,
> And see another's danger, safe at land.' [1]

Of course Lucretius hastens to explain that this is 'not because it is delightful or a pleasure at all that any one should be in distress, but because it is sweet to see dangers from which you yourself are free. It is sweet, too, to see great armies arrayed on the plains struggling in combat without yourself sharing in the danger. But,' Lucretius continues, 'nothing is more pleasant than to occupy the calm high places of philosophy, that are well defended by the learning of the wise, from which you may look down and see others, wandering hither and thither, and going far astray in their search for the way of life, the contest of intellect, the rivalry of rank, the striving night and day with exceeding toil to struggle to the height of power, and be masters of the world. O, wretched minds of man! O, blind souls! not to see in what darkness of life and in how great dangers is this little term of life spent, not to see that nature demands nothing else than for the body to be free from pain, and the mind to enjoy a sense of pleasure free from care and fear.' Of course the 'way of life' is that pointed out by Epicurus.

The pleasure described in the first lines of this passage is a

[1] Creech.

somewhat selfish one. It does, indeed, stir the imagination to behold danger from a place of safety far away; but it is only a cowardly, sentimental soul that can actually enjoy the sight of danger that it would not face itself. Lucretius, we are convinced, would rather have plunged into the waters to save a life at the cost of his own, than stand passive, to enjoy a thrill of poetic sensation at the cost of drowning men. Lord Bacon was unfair in naming this the 'Lucretian pleasure.' Rather are these the words of one who cries to others still in the storm, 'I have found the shelter—come!' Lucretius does but use this as an illustration from which he may pass to the bold figure of the mountain-tops on which the Epicurean stands. There is something very characteristic in the next lines. Sometimes we hear much the same language in our own day from men who have found for themselves new opinions as to God and Hereafter,—who boast a new creed different from the common creed of men. Occasionally they look down on the belief of the many with just such a calm and confident disdain as this; but their hearts are not warm enough for the pity which in Lucretius quite overpowers the disdain. With such a creed as Lucretius professed to have found for himself, and with his fervent temper, he must have felt that the mountain-tops, though lofty places of view, were very cold sometimes. Yet the rarity of their air could not chill the feeling for humanity in his heart.

Afterwards, in some splendid pictures, Lucretius proceeds to show how little wealth or birth or kingly power can deliver men from care and fear. Reason alone can do this. But all this time the atoms have been waiting, and, with a *Nunc age*, Lucretius recalls his reader to the subject.

The Second Book contains, as Professor Jenkin remarks, what may be called the kinetics of the Atomic theory. In it Lucretius promises to treat of the *motion* of the atoms. He will set forth 'by what motion the birth-giving atoms beget different things, and after they are begotten break them up again, and by what force they are compelled to do this, and what swiftness in moving through the void they possess.'[1] The book opens with the proposition that matter does not 'cohere inseparably massed together.'[2] It is always in motion—coming and going. This he infers from the continual change in the world, by which individuals alter and perish while yet the whole remains the same. The cause of these changes, what we should call the energy of the universe, Lucretius holds to be the atoms in motion. The only ultimate form of energy which Lucretius recognizes is the motion of the atoms. His next proposition is to the effect that the atoms can never stop. 'No rest is given to the bodies of the first-beginnings.'[3] After they have come into collision with one another, they cannot either come to a stop or move more slowly,—they rebound in opposite directions, keeping their original velocity. 'For when, being in motion, they have met and clashed, as they so often do, it happens that they suddenly leap asunder in different directions, and no wonder, since they are very hard and of strength proportioned to their solidity,[4] and nothing behind gets in their way.' In this it is of course implied that the atoms are elastic. Professor Jenkin has criticised Lucretius very acutely here. He shows that if the atoms were not elastic, 'they must gradually slacken speed after striking and rebounding, stop

[1] Quo motu genitalia materiai
corpora res varias gignant genitasque resolvant,
et qua vi facere id cogantur, quaeque sit ollis
reddita mobilitas magnum per inane meandi. ii. 62-5.

[2] nam certe non inter se stipata cohaeret
materies. ii. 67-8. Cf. i. 340-5.

[3] ii. 95-6. So Diog. L. x. 43, κινοῦνται συνεχῶς.

[4] Thus Mr. Munro renders,
nam cum cita saepe
obvia conflixere, fit ut diversa repente
dissiliant; neque enim mirum, durissima quae sint
ponderibus solidis neque quicquam a tergo ibus obstet.
ii. 85-8.

for an inconceivably short time, and then gradually resume their pace in an opposite direction.' If they rebound, before moving on again they must stop. Modern science explains that, even if they do stop, their energy yet remains unchanged, for the former energy of motion is now transformed into heat, vibration, or some other form of energy. It will be remembered that Lucretius's atoms have no secondary properties, but only hardness and, as he assumes, elasticity. But in a perfectly hard body such as he conceives, motion cannot be transformed into heat or anything else. We now know that a body which is perfectly hard is not elastic. Lucretius did not know this. His atoms must have come to a stop, and this 'would be equivalent to the destruction of matter.' The next proposition has been anticipated at the end of the First Book, where it is rather implied than actually stated. It is that the atoms, as combined in various bodies, are in motion; they 'mutually give and receive motions.'[1] For some reason or other Lucretius thinks it hardly necessary to state this as a dogma by itself,—probably because he views it as implied in the last proposition. It is indeed constantly referred to and implied throughout the poem. He proceeds at once to defend it.[2] 'You need not marvel at this, why, seeing that the first-beginnings of things are all in motion, still the sum appears to stand in perfect rest.' The atoms of any body may move to and fro ceaselessly while we can see nothing but a mass of matter in repose. Just so, a great flock of sheep and lambs, all cropping the grass or gambolling on it, when seen from a distance appears to be only 'a white spot standing on the green hill,' or a mighty army of foot and horse, all in rapid motion,

[1] Atque eadem magni refert primordia saepe
cum quibus et quali positura contineantur,
et quos inter se dent motus accipiantque.

i. 817-19; repeated at 908-10.

[2] ii. 308-32.

if seen from some place on the mountains very far away, appears only a small bright patch at rest on the plain. Even so, the smallness of the atoms puts, as it were, a vast distance between their motions and our eye.

Lucretius further develops this statement as accounting for the different densities of various bodies. In some bodies the atoms rebound, leaving smaller intervals ; in others they leave larger. In a mass of iron or stone, the atoms are entangled with one another, and can only throb or oscillate, moving to and fro within very small distances ; in softer bodies, like air or sunlight, the atoms rebound at greater intervals.[1] We gather also as a deduction from the last proposition that the atoms, even when they form such a mass of stone or iron, still move as swiftly as they did when streaming through the void. If some rebound within very small limits, they must move to and fro oftener than those which form more porous bodies. (The modern explanation of density, of course, is not merely more molecules within a given space, but perhaps molecules of greater weight also.)

It is worth while to pause for a moment to think how remarkable this statement of Lucretius is. A lump of stone or iron certainly does not give to our senses any impression that its particles are in motion: the piece of inert matter certainly *appears* to be at rest. It is not easy to see what could have

[1] sed magis adsiduo varioque exercita motu
partim intervallis magnis confulta resultant,
pars etiam brevibus spatiis vexantur ab ictu.
et quaecumque magis condenso conciliatu
exiguis intervallis convecta resultant,
indupedita suis perplexis ipsa figuris,
haec validas saxi radices et fera ferri
corpora constituunt et cetera de genere horum
paucula quae porro magnum per inane vagantur.
cetera dissiliunt longe longeque recursant
in magnis intervallis : haec aera rarum
sufficiunt nobis et splendida lumina solis. ii. 97-108.

suggested to the discoverer a thought so opposite to what the senses tell us. Yet it is accepted by science now as certainly true, both for solid bodies, liquids, and gases. In solids, indeed, these motions of the molecules are confined within very narrow limits, and cannot be detected ; yet Professor Tyndall says of the atoms composing the hardest body, when heated, 'They collide, they recoil, they oscillate.'[1]

According to Maxwell, 'the principal difference between a gas and a liquid seems to be that, in a gas each molecule spends the greater part of its time in describing its free path, and is for a very small portion of its time engaged in encounters with other molecules ; whereas in a liquid the molecule has hardly any free path, and is always in a state of close encounter with other molecules.' In both liquids and gases the molecules move more freely than in solid bodies, and the argument drawn from the diffusion of gases and liquids forms one of the strongest proofs of the motion of molecules. How could two different gases mix so very rapidly, unless the molecules composing them were in motion ? The molecules of any gas flying about beat against whatever opposes them, and the constant succession of these strokes, according to the Atomic theory, explains the pressure of gas. Further, as Maxwell says, 'All the three kinds of diffusion, the diffusion of matter, of momentum, and of energy, are carried on by the motion of the molecules.' Heat, viewed as a mode of motion, furnishes another argument. Lucretius states that the molecules of bodies are moving with more or less speed. Now if heat be a mode of motion of gross matter, then, as all bodies are more or less hot, the molecules of all bodies must be moving more or less quickly. This is just what Lucretius says, and this statement of his is perhaps his most marvellous anticipation of modern scientific discovery.

[1] 'Fragments of Science for Unscientific People,' by John Tyndall. 1871. Page 12.

Lucretius continues with an illustration. If you wish to realize what the motion of the atoms is, observe when the sunlight streams into a dark chamber, how numberless motes toss about and dash against one another, meeting and breaking away again. These may help us to conceive how the atoms are for ever tossing about in the great void. Moreover, in this case a small thing illustrates a great truth. For the movements of the motes have a cause: they imply that 'hidden and invisible motions of the atoms are at the bottom.' The atoms move first of themselves, next they strike against those bodies which are formed of a few atoms in union, and make them move, these again stir up those which are a little larger. 'Thus motion mounts up from the first-beginnings and step by step issues forth to our senses.'[1]

Lucretius next points out that the velocity of the atoms passing through the void is immense. Notice, he says, at sunrise,—an Italian sunrise, we must remember,—after the first rays have begun to shoot and the birds to sing in the woods, how soon and how suddenly the heaven is filled with light. Yet the rays of light are formed of countless molecules, and have to pass through a medium, the air, the molecules being pulled back by each other and hindered by the air. How much more swiftly must the atoms, which are single bodies, stream through the unresisting void? Professor Jenkin remarks that Lucretius 'may also have felt that if all the power of the universe depended on the motion of exceedingly small particles, it was necessary to suppose them endowed with great velocity; but we do not find this argument used, although it has led the modern believer in atoms to the conviction that, if their motion does represent energy, their velocity must be

[1] ii. 114-41. The tumblings of the motes which our eyes clearly discern, so to say, make visible to us the invisible plagae of the atoms to which they are due (ll. 134-6). Lucretius evidently feels this illustration to have scientific importance.

enormous. Lucretius would be glad to know that Herapath, Joule, Krönig, Clausius, and Clerk-Maxwell have been able to calculate it.'

Dr. Joule calculated the actual velocity of the molecules of hydrogen, and found it to be exceedingly great, at the rate of nearly sixty-nine miles a minute. The velocity of other gases is less. Maxwell has calculated, from the data of Professor Loschmidt of Vienna, the actual velocity of the molecules of four different gases at 0°C. It is as follows:

	Hydrogen.	Oxygen.	Carbonic oxide.	Carbonic acid.
Metres per second . . .	1,859	465	497	396

The molecules of calm air, he says, are flying about in all directions at the rate of about seventeen miles a minute:

'If all these molecules were flying in the same direction, they would constitute a wind blowing at the rate of seventeen miles a minute; and the only wind which approaches this velocity is that which proceeds from the mouth of a cannon. How, then, are you and I able to stand here? Only because the molecules happen to be flying in different directions.'

'But it is not only against us, or against the walls of the room, that the molecules are striking. Consider the immense number of them, and the fact that they are flying in every possible direction, and you will see that they cannot avoid striking each other. Every time that two molecules come into collision, the paths of both are changed, and they go off in new directions. Thus each molecule is continually getting its course altered, so that, in spite of its great velocity, it may be a long time before it reaches any great distance from the point at which it set out.'

Again, referring to an experiment with ammonia, he says:

'The molecules of ammonia have a velocity of 600 metres per second, so that if their course had not been interrupted by striking against the molecules of air in the hall, every one in the most distant gallery would have smelt ammonia before I was able to pronounce the name of the gas. But, instead of going at this rate, each molecule of ammonia is so jostled about by the molecules of air, that it is sometimes going one way and sometimes another. It is like a hare which is always doubling, and though it goes at a great pace, it makes very little progress.'[1]

Maxwell has calculated also the number of collisions which each molecule must undergo in a second. They amount to thousands of millions, and are as follows:

	Hydrogen.	Oxygen.	Carbonic oxide.	Carbonic acid.
Collisions in a second (millions)	17,750	7,646	9,489	9,720

'No wonder,' he observes, 'that the travelling power of the swiftest molecule is but small, when its course is completely changed thousands of millions of times in a second.' So circumstantially has science developed the thought of Lucretius that the atoms really move with as great velocity when pent in stone as when floating free in the void. In the same lecture Maxwell divides the ultimate results of molecular science into three ranks, 'according to the completeness of our knowledge of them.' In the first rank he places the relative masses and the velocities of the molecules, which, he says, 'are known with a high degree of precision.' Other data, which are less precise, he places in the second rank, and others which are, as yet, conjectural, in the third. Astonishing, indeed, is this weighing of the atoms and counting their numbers and measuring their speed,—to realize how, after long and patient processes of thought, the atom, hunted in the dark till it has been found, appears at length before the eye of the investigator. Indeed, a triumph of the 'scientific imagination' and intellect! Yet among the propositions of Lucretius, which have been stated, the fruit of ancient thought unaided by experiment, there are one or two which seem to us even as wonderful.

Several additional propositions may be gathered from the poem, which, though not stated in order by Lucretius, may be here inserted. The first of these is not distinctly formulated, but is often referred to or implied throughout. Lucretius

[1] 'Molecules.' A Lecture by Professor Clerk-Maxwell. 1873.

applies to the combination of the atoms before Matter is produced, a word which is most unusual as employed to denote things without life, namely, *concilium*. No editor has yet explained its exact force as a specific scientific term in Lucretius's poem. Apparently it must have conveyed to a Roman ear the meaning of 'an assemblage' of living beings, according to its ordinary sense. Lucretius repeats again and again that before any kind of substance (res), whether dead-matter or living organism, can be formed, atoms must enter into concilium. In some passages Lucretius seems to use the word where we should speak of a germ (of organic) or a nucleus (of inorganic matter). When the tossing of the sea of matter has driven together atoms of like kind,[1] these come close together so as to continue to rebound at the shortest possible distances. A small body is thus formed, and, when other atoms have joined it, it increases in bulk, and so any kind of substance is produced. On the other hand, if unlike atoms come together, they are unable to assimilate their motions, and thus, by continuing united, to form a nucleus. In such cases, though atoms are for the time in close union, concilium does not occur.[2] Lucretius supposes every

[1] Cf. ii. 563-4. If any kind of atoms are finite in number, they will never be able to combine,

> nunquam in concilium ut possint compulsa coire,
> nec remorari in concilio, nec crescere adaucta.

[2]
> multaque praeterea magnum per inane moventur,
> conciliis rerum quae sunt reiecta nec usquam
> consociare etiam motus potuere recepta. ii. 110-12.

> et quaecumque magis condenso conciliatu
> exiguis intervallis convecta resultant
> indupedita suis perplexis ipsa figuris,
> haec validas saxi radices et fera ferri
> corpora constituunt. ii. 100-4.

> propter dissimilis formas variasque figuras,
> quod non omnia sic poterant coniuncta manere
> nec motus inter sese dare convenientis. v. 440-2.

We may contrast with these the lines immediately preceding the last kind of matter to be formed by the growth of such nuclei. Unless atoms are able thus to move in concert and to continue so moving, no body can be produced. As to the conception of atoms united in concilium acting as a kind of nucleus, we may quote Professor Clifford, who says, 'It is known that the first step is often a difficulty in the formation of chemical compounds, and that when the process has once begun, the new compound has the property of assisting the formation of its like.'[1]

Professor Veitch[2] has said that 'atomic motion of the kind supposed (by Lucretius) would give at best only aggregates of particles.' The very same criticism was made long ago by Plutarch,[3] who complains that the Epicurean atoms could compose nothing, but could only make 'many and continual collisions,' and that the constant clashing of the atoms produces 'no incorporation or coalition, but only percussions and re-percussions'[4]—only 'a confusion and combat of atoms, which the Epicureans choose to call generation.'[5]

passage. They refer to atoms, which, being unlike, are unable to combine.

nec similis nostris rebus res ulla videri,
sed nova tempestas quaedam molesque coorta
omne genus de principiis, discordia quorum
intervalla, vias, conexus, pondera, plagas,
concursus, motus turbabat, proelia miscens,
propter dissimilis formas variasque figuras, &c. v. 435-40.

[1] 'Lectures and Essays,' by Professor Clifford, vol. ii., p. 312.

[2] 'Lucretius,' &c., p. 39.

[3] Against Colotes the Epicurean, c. 10.

[4] *πληγὰς καὶ ἀποδόσεις.*

[5] To exactly the same effect says Simplikius, 'De Coelo,' p. 133 (ed. Karsten, 1865):—The eternal atomic collisions may produce entanglement and mechanical contact, but no fusion, into one new nature. *συμψαύειν μὲν αὐτὰ καὶ πλησίον ἀλλήλων εἶναι ποιεῖ· φύσιν μέντοι μίαν ἐξ ἐκείνων κατ' ἀλήθειαν οὐδ' ἡντιναοῦν γεννᾷ . . . . τοῦ δὲ συμμένειν τὰς οὐσίας μετ' ἀλλήλων μεχρὶ τινὸς αἰτιᾶται τὰς ἐπαλλαγὰς καὶ τὰς ἀντιλήψεις τῶν σωμάτων. τὰ μὲν γὰρ αὐτῶν εἶναι σκαληνὰ, τὰ δὲ ἀγκιστρώδη,* &c.

Lucretius, though he may not have provided for anything more than this, according to modern scientific requirements, would certainly not have agreed to this definition as correct. He insists pointedly that the atoms must unite *in concilio,* that is to say, the atoms must combine and group themselves in a certain way before any kind of substance can be produced. Thus he does certainly insist on something more than mere mechanical combination among the atoms. According to Lucretius, concilium occurs only when like atoms come together.[1] Modern chemistry, too, teaches us that all the atoms of each element are alike, while the atom of each elementary substance differs essentially from that of every other. Chemistry, moreover, asserts that both simple and compound bodies are composed of molecules, or groups of atoms, which are formed by the action of chemical affinity. Can it be said that by his notion of concilium, Lucretius meant to assert that no kind of substance can be produced unless the atoms have first united in such order as to form the molecules of that substance? and that concilium thus foreshadows the doctrines of *chemical combination* among the atoms, and of the *molecular composition of matter?* The existence of molecules is now viewed not as contradictory to the theory of atoms, but only as its complement, and it would be interesting to find in Lucretius a foreshadowing, though faint, of the doctrine of molecules developing out of the atomic theory as held by Epicurus.

[1] We may compare the passage of Hippolytus on the atomistic origin of the world. Its germ is formed in consequence of like atoms being thrown together. *προσκρόυοντα ἀλλήλοις συμπλέκεσθαι τὰ ὁμοιοσχήμονα καὶ παραπλήσια τὰς μορφαις.* He is here speaking of Leucippus ('Refut.,' i. 12). Again, Simplikius, 'De Coelo,' p. 110 (Karsten), of Democritus's account of the origin of the world, *τὰς ἀτόμους . . . συγκρούεσθαι, καὶ τὰς μὲν ἀποπάλλεσθαι ὅποι ἂν τύχωσι, τὰς δὲ περιπλέκεσθαι κατὰ τὴν τῶν μεγεθῶν καὶ σχημάτων καὶ θέσεων καὶ τάξεων συμμετρίαν.* Democritus insisted specially on the law that like atoms are attracted to each other. He finds the cause of this in the whirling motion which, as it were, 'sifts' the atoms into separate heaps, according to their size and form. (See our note on chap. iv. § 3.)

Three other propositions are introduced by Lucretius only in the later books. One very important one may be called the doctrine of Emanations. From the surface of all bodies minute particles are at every moment streaming off. These particles may preserve the shapes of the bodies which discharge them, and so form images or 'idols,' which are the cause of our seeing. Such effluxes cause us either to see or hear or smell, according as they enter the passages appropriate to each sense. There is a constant succession of these emanations. 'In one moment of time there must be carried away from objects idols many in number, and in many ways, and in all directions round.'[1] If these images were at any moment to cease, we should also cease to see or hear or smell. This theory of emanations is one of the most distinctive of all Epicurus's doctrines. In consequence of this constant evaporation from all matter, a perpetual waste must be going on in the world, and this helps us to see why Lucretius thought it so necessary that the world should be constantly fed with fresh matter from without in order to preserve it in being.

Again, all bodies are porous, and so these emanations constantly fly through them in all directions. Thus all matter is more or less interpenetrated by other matter. Every substance, at any moment, contains within its pores particles of many other substances. Though Lucretius does not state it so prominently as Empedocles did, the notion of pores resembling or differing in form from the shapes of the atoms, plays an active part in his system. Lucretius falls back on the Empedoclean doctrine of the *συμμετρία τῶν πόρων*, that is, the symmetry between the pores and the shapes of the atoms entering them, as an explanation of not a few questions.

Moreover, the pores of all bodies are filled with air. The air that is within each substance is in constant motion, throbbing to and fro, and making the whole body vibrate. In explaining

[1] iv. 163-5.

the operation of the magnet, he says, 'All things must have air in their body. . . . This air, therefore, which is hid in the inmost parts of the iron, is constantly throbbing in restless motion ; and therefore, without doubt, it beats upon the ring, and must stir it inwardly.'[1] Lucretius works out at great length the doctrine of Images, as explaining the faculty of sight, in his Fourth Book. All three doctrines, however, are pointedly illustrated in his explanation of the magnet,[2] which we discuss in a later chapter. They show us how Epicurean science conceived all matter to be penetrated by invisible movements of many various kinds, intricately crossing each other.

Before setting forth what the original motion of the atoms is, Lucretius shows that there is no exception to the universal tendency of gravitation. All motion, relatively to the earth, is downwards. The direction of the atoms is also downwards. Borne by their weight, they fall 'straight down' in infinite numbers through infinite space with immense speed. We can fancy the bewildering vision of the falling atoms haunting the poet's mind both awake and in his dreams. Lucretius having now arrived at atoms in motion, sees the way clearly to construct the world.

His next proposition is a most remarkable one. In it he suddenly deserts the domain of physics altogether. *Illud in his rebus*, Lucretius begins,—a rather prosaic phrase, of which he is fond when he wishes to call special attention to some point of the argument :

[1]
denique res omnes debent in corpore habere
aera, quandoquidem raro sunt corpore et aer
omnibus est rebus circumdatus adpositusque.
hic igitur, penitus qui in ferrost abditus aer,
sollicito motu semper iactatur, eoque
verberat anellum dubio procul et ciet intus
scilicet. vi. 1034-40.

[2] vi. 906-1064.

'This point of the subject we desire you to apprehend, that when atoms are borne straight downwards through the void by their own weights, at quite uncertain times and uncertain places, they push themselves a little from their course, only just so much that you can call it a change of inclination. If they were not wont to swerve thus, they would fall down all, like drops of rain, through the deep void, and no clashing could have been begotten, nor any collision produced, among the first-beginnings: thus Nature never would have produced anything.'[1]

He continues,—If any one believes (as did Democritus) that atoms can come into contact through the heavier falling more quickly and striking the lighter, 'he goes far astray from true reason.'[2] This is no substitute for declination. It would be so, indeed, did the atoms fall through water or air, which offer more resistance to the lighter than they do the heavier. Empty void, on the other hand, cannot offer resistance to anything, and, therefore, things of unequal weight fall through it with equal velocity. Had Lucretius known it, he would have quoted the modern experiment showing the resistance of the atmosphere,—a guinea and a feather dropped from the top of a receiver, exhausted of air, and falling to the foot at the same instant. Again, he defines the amount of this inclination more specially as 'not more than the least possible.' No one, he says, can positively state that falling bodies do not 'swerve at all from a straight line.' According to Epicurus, and, as Munro believes,[3] according to Lucretius also, the result of the collisions

[1] ii. 216-24.

[2] So Aristotle, though he thought void impossible, had pointed out before Epicurus's time, that, in a vacuum such as Democritus assumes, falling bodies would experience no resistance, and must all fall at the same rate. It is characteristic of Epicurus that he adopts Aristotle's correction of Democritus.

[3] See his notes on Bk. I., ll. 1000-1,

> partibus e cunctis *inferna*que suppeditantur
> ex infinito cita corpora materiai

and on l. 1044. Cf. also ll. 1035-6,

among the atoms produced by this declination is that they are forced in an upward direction, whether perpendicularly or obliquely upwards.[1] 'Both Epicurus and Lucretius conceived the rising up of the atoms in a direction more or less contrary to the only natural motion, as that which enabled things to come into being and remain in being.' As the natural motion of the atoms is downwards, every kind of upward motion would have an upholding power, and enable things to maintain their existence. 'This swerving,' says Professor Jenkin, 'seems but a silly fancy, and yet consider this: it is a principle of mechanics that a force acting at right angles to the direction in which a body is moving does no work, although it may continually and continuously alter the direction in which the body moves. No power, no energy, is required to deflect a bullet from its path, provided the deflecting force acts always at right angles to that path—an apparent paradox, which is, nevertheless, quite true and familiar to the engineer. It is clear to us that Epicurus, when he devised his doctrine of a little swerving from the straight path of an atom, had an imperfect perception of this mechanical doctrine ; a little swerving would bring his

quod nullo facerent pacto nisi materiai
ex infinito *suboriri* copia posset

and ll. 1049-53, beginning

quare etiam atque etiam *suboriri* multa necessest.

Mr. Munro has thoroughly defended his reading inferna, and his explanation of a point misunderstood by previous editors. He quotes Epicurus (Diog. L., x. 41), *τὰ ὑπερείδοντα καὶ στέλλοντα κατὰ τὰς ἀνακοπάς*, which means, 'the particles supporting from below and rising upwards in consequence of collisions.' We find in Plutarch a passage to exactly the same effect, where the upward motion of the atoms is attributed to 'collision' and 'recoil.' *τὰ δ' ἄνω κινοῦμενα, κατὰ πληγὴν, κατὰ παλμὸν* ('De Plac. Phil.,' i. 12). Lachmann's explanation is consistent neither with the other points of Lucretius's system, nor with the texts of Epicurus.

[1] The declination itself is not in an upward direction. For a falling body to move up is rather a violent alteration of its course.

atoms into contact, and a modern mechanician would tell him you require no power to make them swerve.' It may be so. The Greek mind had marvellous intuitions. An observant man could hardly have failed to notice that exceedingly little force is required slightly to deflect from its path a thing which is already in motion. Yet we can hardly think that Epicurus had such an idea in his mind: was it not rather an exceedingly simple, yet most original, solution of the difficulty,—given an infinite number of atoms, moving all in parallel lines, falling straight downwards and never touching one another, how to make them meet and combine, that they may create the world? This least possible declination, so little that it was hardly moving from the straight line, was sufficient and answered every need. There could be no simpler solution of a difficulty.

Professor Jenkin finds two inconsistencies in this part of Lucretius's theory. Firstly, as to the downward motion of the atoms, it is plain that 'Lucretius unconsciously assumed the world as his basis by which to measure direction and velocity.' This objection is one which has been made by many critics of Epicurus, and appears unanswerable. But it is clear that by *up* and *down* Epicurus simply denotes the two opposite directions of motion in infinite space, so that the notion of a downward motion of the atoms is merely relative in its meaning.[1] The second objection (which Munro has quoted, appa-

[1] 'From the time of Cicero [see also Plutarch, "De Defectu Orac.," c. 28], Epicurus has been reproached with his simplicity in admitting a downward motion, and, consequently, an *up* and a *down* in infinite space. But a text of Epicurus's definitely proves that he was not so simple. Up and down are conventional terms, which designate, according to him, the two opposite directions of movement in infinite space. *ὥστ' ἔστι μίαν λαβεῖν φοράν, τὴν ἄνω νοουμένην ἐις ἄπειρον, καὶ μίαν τὴν κάτω, ἂν καὶ μυριάκις πρὸς τοὺς πόδας τῶν ἐπάνω τὸ παρ' ἡμῶν φερόμενον ἐπὶ τοὺς ὑπὲρ κεφαλῆς ἡμῶν τόπους ἀφικνῆται, ἢ ἐπὶ τήν κεφαλὴν τῶν ὑποκάτω τὸ παρ' ἡμῶν κάτω φερόμενον. ἡ γὰρ ὅλη φορὰ οὐδὲν ἧττον ἑκατέρα ἑκατέρᾳ ἀντικειμένη ἐπ' ἄπειρον νοεῖται.* (Diog. L., x. 60.) Thus *up* and *down* denote for Epicurus a relative condition, like the terms right and left, bass and treble, large and small.

rently with approval) is a more important one. We have stated that Lucretius sought for an explanation of the power of the universe in the velocity of his atoms. But atoms pouring downward all at one speed, and in parallel lines, could really be no source of power. 'Motion in mechanics has no meaning except as denoting a change of relative position.' But it would be impossible for these atoms ever to change their relative position; they would never be nearer, and could never be more distant from each other than before. Lucretius's atoms are all, relatively to one another, perfectly still and motionless. 'Atoms pouring onward, as imagined by our author, could be no source of power.' This fact, taken by itself, is of course undeniable. Again, he says of the atoms of Democritus (which moved to and fro in all directions indifferently), 'one atom might then exert its force on another, but the Lucretian atoms would have remained in profound stillness, *except for that occasional swerve.*' But are all these remarks justified? Lucretius, of course, saw that his atoms, in their original downward movement, were relatively motionless. No collision could then take place among them. The truth is that Lucretius never conceived an atom as acting on another in any way beyond the declination and but for it. Professor Jenkin writes as if Lucretius's theory assumed or implied that the atoms could combine or act on one another even without declination. Now, surely, Lucretius[1] states emphatically that, but for declination, the atoms would never have touched each other, and 'nature never have produced

What remains always untenable is the hypothesis which attributes to original motion *two* possible directions and not more.'—'La Morale d'Épicure,' pp. 74-5. Epicurus's defence is, that while there can be no absolute Above and Below in infinite space, still motion from our head towards our feet must always be contrary to motion from our feet towards our head. This seems reasonable enough, though Zeller will not allow it. Vol. ii., p. 241, note 2 (Eng. tr.).

[1] At ll. 216-24.

aught.' The number of atoms being infinite, Lucretius saw that the slightest declension must produce innumerable collisions. In these collisions, of course, the whole velocity of the atoms comes into action, and they thus develop an ample 'source of power.' The Lucretian atoms are driven about, and move to and fro, even more freely than did those of Democritus. The illustration of the motes in the sunbeam is used to express faintly with what restless and promiscuous motion they dash about. The swerve *does* universally change their relative position. This is just the force of it, and Lucretius (whether reasonably or not) thought this quite sufficient as a means to bring his atoms into contact. Whether it be sufficient or not, he is not inconsistent with himself in this. The slightest declension is quite sufficient to bring his atoms into universal collision, and so to render them a source of power, the source of all the energy or forces acting in the world.

But Lucretius has a double purpose in this remarkable doctrine. First, the mere fact of swerving is enough to bring his atoms into contact. In the second place, by this it is intended to explain how we come to possess free-will. If the atoms could not decline, neither should we be able to move at will.

This power in the atoms corresponds to free-will action in men and animals ; and in the conception of Lucretius it is the cause of it. Professor Jenkin here suggests the alternative that, instead of permitting atoms to deflect their path at will, Epicurus might have given to man the power of deflecting the stream of circumstance. He says forcibly,—

'The atoms may, as Democritus believed, build up a huge mechanical structure, each wheel of which drives its neighbour in one long inevitable sequence of causation ; but you may assume that beyond this ever-grinding wheelwork there exists a power not subject to, but partly master of, the machine ; you may believe that man possesses such a power, and if so, no better conception of the manner of its action could be devised than the idea of its deflecting the atoms in their onward path to the right or left of that line in which they would naturally move. The will, if it so acted, would add nothing sensible to, nor take anything sensible from, the energy of the universe. The modern believer in free-will will probably adopt this view, which is certainly consistent with observation, although not proved by it. Such a power of moulding circumstances, of turning the torrent to the right, where it shall fertilize; or to the left, where it shall overwhelm; but in nowise of arresting the torrent, adding nothing to it, taking nothing from it—such is precisely the apparent action of man's will.' [1]

Epicurus, accepting the Atomic theory with all its assumed consequences, thought himself compelled from his point of view, either to accept Necessity (as Democritus had done) or to endow his atoms with Free-will, exercised not constantly, but at uncertain intervals. It is not every one who would have thought of freeing himself from a difficulty in such a way, or would have had the courage to assign free-will to atoms.

The latest historian of Epicureanism has called this doctrine of atomic declination 'the central and most original point of

[1] This passage from Professor Jenkin suggests to us the question, How far does his expression as to the action of free-will hold for the moral world? With regard to different motives for personal conduct, the action of free-will appears, similarly, to be *rarely* (in the case of most people, at least) in violent opposition to that of impulse, but as it were slightly, yet distinctly, deflecting the spontaneous impulse. In illustration of this we quote from an able Roman Catholic writer: 'I have the fullest power of *opposing* my will's spontaneous impulse. . . . How far I may *choose* to put forth such exertion,—this is not abstractedly matter of calculation at all. . . . At the same time, it should be observed that, in all ordinary cases, the act of will, which results in fact, is found in *close vicinity* to the will's spontaneous impulse. . . . In 999 cases out of 1,000 a man's probation is carried to a successful issue by this more than by anything else; viz., by putting forward on repeated occasions a number of acts, which are *a little* higher than his spontaneous impulse' (Dr. W. G. Ward, 'Science, Prayer, Free Will, and Miracles,' 1881, pp. 46-7). In great crises of personal history, involving moral and spiritual struggles, certainly this does not apply; but for the ordinary life of most people it is, probably, true enough.

Epicurus's system.'[1] In order the better to understand it, we must leave it for the present until we have examined Epicurean psychology.

The last proposition which we shall quote from Lucretius is as follows:—'Nor was the store of matter ever more dense or ever separated by larger intervals than now; for nothing is either added to it or lost from it. Wherefore, in whatsoever motion the bodies of the first-beginnings now move, in the same way they moved in time gone by, and in like manner will they always be borne along hereafter. . . . No force can alter the sum of things [*i.e.* the universe]; for there is nothing outside into which either any kind of matter can escape out of the universe, or out of which a new force[2] can arise and burst in and change all the nature of things and disturb its motions.'[3] Lucretius here states that matter was never more or less dense than it is now, and that the atoms have always moved, and always will move, with the same velocity; and because there is nothing else outside and beyond the atoms, nothing can alter

[1] M. Guyau, 'La Morale d'Épicure,' 1881, p. 99, note.

[2] Nova vis. Munro translates 'a new supply'—that is, of matter.

[3] Nec stipata magis fuit unquam materiai
copia, nec porro maioribus intervallis;
nam neque adaugescit quicquam neque deperit inde.
quapropter quo nunc in motu principiorum
corpora sunt, in eodem ante acta aetate fuere
et post haec semper simili ratione ferentur,
et quae consuerint gigni gignentur eadem
condicione et erunt et crescent vique valebunt,
quantum cuique datum est per foedera naturai.
nec rerum summam commutare ulla potest vis;
nam neque, quo possit genus ullum materiai
effugere ex omni, quicquam est extra, neque in omne
unde coorta queat nova vis inrumpere et omnem
naturam rerum mutare et vertere motus. ii. 294-307.

We may compare v. 361 ff,

sicut summarum summa est aeterna neque extra
qui locus est, quo dissiliant, neque corpora sunt quae
possint incidere.

the sum of things, 'what we should call the energy of the universe.' 'This proposition,' says Professor Jenkin, 'foreshadows the doctrine of Conservation of Energy. It is clear that Lucretius conceived two things as quite constant: atoms were neither created nor destroyed, and their motion could neither be created nor destroyed. He believed that each atom kept its velocity unaltered. The modern doctrine is that the total energy of the universe is constant, but may be variously distributed, and is possibly due to motion alone ultimately, though this last point has not been yet proved.' 'If matter in motion be conceived as the sole ultimate form of energy, *Lucretius must be allowed great merit in having taught that the motion of matter was as indestructible as its material existence*, although he knew neither the laws of momentum nor of *vis viva*. If energy, as he believed, be due solely to motion, then his doctrine is true.' Thus in the concluding proposition, Lucretius states that Force is indestructible.

Lucretius has now constructed a complete atomic theory of the world. He has now completed the needful foundation from which to prove that the atoms, moving of their own accord, have created all existing things, including the body and soul of Titus Lucretius and of all men destined to be. If you are not convinced of this,—if you are not at once prepared to do homage to the creative atoms, the poet, like Professor Clifford, waxes indignant, and wonders that you are so impervious to reason.

We have now enumerated the whole series of propositions containing Lucretius's Atomic theory. His scientific style is admirably simple; its simplicity and plainness convey the impression of good faith. We shall refer later on to the illustrations which break the severity of the argument. Apart from their beauty of conception, they have, at the same time, in every case, a scientific value: they show, according to Tyndall, that Lucretius had a strong 'scientific imagination.'

CHAPTER IV.

THE BIRTH OF THE WORLD.

THUS far Lucretius has carried us with wonderful coherence. But, after following him as our guide so far, we now come to a gap in his theory, indeed a bottomless chasm over which he has thrown no bridge. With a single leap he passes from the whirling atoms to the world with all its life, beauty, and order, but hardly a word as to *how* the atoms have produced it, how the supreme result is reached. Sellar says: 'He may, as was natural, have failed in adequately conceiving the transition from the fortuitous concourse of lifeless atoms to the exuberant life and perfect order of the world:' perhaps it might be more correct to say—he almost totally omits any attempt to show how this could take place. Over and over again he asserts that the intricate and countless movements of the clashing atoms, trying every manner of combination in the course of their perpetual motion from eternity, have produced the world, and that all the life upon it has resulted from the complicated motions and collisions of these hard little kernels. But for sole answer to the question, 'How can this take place?' Lucretius gives a few vague hints. 'Truly not by design have the first-beginnings of things stationed themselves each in their proper places, by sage consideration, nor have they made agreement what motions they should each assume. Not so in truth; the cause is that they are many in number, and have shifted, in changes many, all the universe over. They have been driven together and tormented by constant shocks from all eternity. After trying in this way motions and unions of every kind, they fall at length into the arrangements out of which this world of ours has been formed, and by which too it has been preserved in being through many cycles, when once it has been thrown into the fitting motions.'[1] This passage contains Lucretius's whole account of Evolution, certainly a short one. He repeats it often, and frequently enlarges it by some phrase like the following:—'At length come together those atoms which, *being suddenly* carried together, become the rudiments' of the world, or 'at length there filtered through those atoms which, *being suddenly cast together*, become, from time to time, the rudiments of great things, of earth, sea, heaven, and of the race of living things.'[2] Lucretius holds that there is something sudden in this atomic begetting of an infant world. When the course of atomic combinations has at last brought the fit atoms near each other, suddenly they leap together, and the germ of the world is born.

[1] i. 1021-30. The same passage, with slight variations, is found at ii. 1053-63; v. 187-94; and, with fullest detail, at v. 419-31, as follows:—

> nam certe neque consilio primordia rerum
> ordine se suo quaeque sagaci mente locarunt,
> nec quos quaeque darent motus pepigere profecto,
> sed quia multa modis multis primordia rerum,
> ex infinito iam tempore percita plagis
> ponderibusque suis consuerunt concita ferri
> omnimodisque coire atque omnia pertemptare,
> quaecumque inter se possent congressa creare,
> propterea fit uti magnum volgata per aevom
> omne genus coetus et motus experiundo
> tandem conveniant ea quae convecta repente
> magnarum rerum fiunt exordia saepe,
> terrai maris et coeli generisque animantum.

[2] Tandem conveniant ea quae convecta repente. v. 429.
> tandem colarunt ea quae, coniecta repente,
> magnarum rerum fierent exordia semper. ii. 1061-2.

Cf. also i. 1030,

> ut semel in motus coniectast convenientis.

As to the variety in nature, the fact that all the individuals of the same kind differ in appearance, Lucretius does endeavour in a vague way to account for it. In the first place, the atoms are not all the same, but vary in form ; and things which differ from one another are composed of atoms of unlike shape. Secondly, the atoms admit of many modes of combination. Lucretius often repeats the formula,—'It matters much with what others and in what position the same atoms are severally held in union, and what motions they mutually give and receive.'[1] Its meaning is that the differences between all bodies are accounted for by differences in the mutual relations of the atoms. They differ in their

> 'Intervalla, vias,[2] conexus, pondera, plagas,
> Concursus, motus,'
> 'Concursus, motus, ordo, positura, figurae,'

'in the spaces between them, their passages, manners of being linked together, weights, collisions, clashings, motions,' and also 'in their arrangement, position, and shapes.' By their differences in shape, motion, weight, and arrangement, the various degrees of colour, sound, scent are produced. To illustrate this he uses the letters of the alphabet. Just as the same letters in different arrangements produce words of entirely different meaning, so the same atoms, in different relations as to order, motion, &c., or with some small exchange for atoms of different form, may produce things of quite opposite qualities, such as fire and air. 'There are certain bodies (that is, the atoms[4]) possessed of such a nature that, if they have haply

[1] ii. 1007-9.

[2] 'Intervalla, vias'—that is, the shapes of the pores between the atoms of each substance. Lucretius nowhere treats this subject in detail as we should expect him to do ; but the doctrine that the shapes of the inter-atomic or inter-molecular pores differ in each substance according to its structure, plays an important part in his system.

[3] v. 438-9 and i. 685.

[4] Lucretius is here refuting the notion of elements.

produced fire, the same may, after a few have been taken away and a few added on, and their order and motion changed, produce air; and all other things may in this way interchange with one another.'[1] 'Nay, even in our own verses,' he says, 'it matters much with what other letters, and in what order, the several letters are placed. If all the letters are not entirely alike, still by far the greatest part are; but the words which they compose differ through the position of these letters. Thus, in material things as well, when the clashings, motions, arrangement, position, and shapes of matter [*i.e.* of the atoms] change about, the things must change too.'[2] Only as there are far more different kinds of atoms than there are of letters, the former can unite in many more combinations, producing different substances.[3] Thirdly, everything is composed of more than one kind of atoms. 'There is nothing which is not formed by a mingling of seeds.'[4] The more different properties and powers that anything possesses, the greater number of different

[1] quin potius tali natura praedita quaedam
corpora constituas, ignem si forte crearint,
posse eadem demptis paucis paucisque tributis,
ordine mutato et motu, facere aeris auras,
sic alias aliis rebus mutarier omnis. i. 798-802.

[2] quin etiam refert nostris in versibus ipsis
cum quibus [sc. elementis] et quali sint ordine quaeque locata.
si non omnia sunt, at multo maxima pars est
consimilis; verum positura discrepitant res.
sic ipsis in rebus item iam materiai
concursus, motus, ordo, positura, figurae,
cum permutantur, mutari res quoque debent. ii. 1015-22.

[3] Tantum elementa queunt permutato ordine solo;
at rerum quae sunt primordia, plura adhibere
possunt unde queant variae res quaeque creari. i. 827-9.

[4] nilesse in promptu quorum natura videtur,
quod genere ex uno consistat principiorum,
nec quidquam quod non permixto semine constet. ii. 583-5.

cetera consimili mentis ratione peragrans

shaped atoms must it contain.[1] Again, life apparently depends merely upon the regular continuance of certain movements of the atoms. A blow produces death by altering the positions of the atoms, and 'entirely stopping the vital motions.' In fact, Lucretius conceives life as a 'mode of motion.' Of course, he has no protoplasm to bridge over the gulf between dead atoms and living beings. In fact, it seems never to have entered his mind that any reasonable man should doubt that atoms, if they do exist, moving in the way he described, would in the course of time produce life.

According to Lucretius, difference of atomic structure in any two substances is sufficient to account for any difference in their qualities. Thus, after describing the terrific force of lightning, he considers this amply accounted for by saying

> invenies igitur multarum semina rerum
> corpore celare et varias cohibere figuras. ii. 677-9.

> nimirum quia multa modis communia multis
> multarum rerum in rebus primordia mixta. i. 814-5.

> atque eadem magni refert primordia saepe
> *cum quibus* et quali positura contineantur. i. 817-8.

> namque eadem coelum, mare, terras, flumina, solem
> constituuent, eadem fruges, arbusta, animantis,
> verum *aliis* alioque modo commixta moventur. i. 820-3.

The same is of course implied in the illustration from the letters of the alphabet; see the whole argument of ii. 581-99 and 661-99. Yet does not Lucretius's explanation of lightning as formed, apparently *solely*, of very small round atoms, seem to contradict this law? Or can it be that lightning is not included in this statement? It is difficult to bring this doctrine into harmony with the other points of Lucretius's atomic theory. Yet Democritus also seems to have held the same, and, on the other hand, to have explained the possession of any one quality in special intensity, as caused by such a substance being formed of one kind of atoms only, *e.g.* fire, on account of its extreme mobility, solely of round and smooth atoms; see Zeller, vol. ii., p. 324 (Eng. tr.).

[1] et quodcumque magis vis multas possidet in se
atque potestates, ita plurima principiorum
in sese genera ac varias docet esse figuras. ii. 586-8.

that the thunder-bolt is formed 'of particles especially minute and ready to move.'[1] Similarly, the mind is formed of exceedingly small, smooth, and round atoms. Lucretius feels that if one thing is made of small, smooth, and round atoms, and another of large, rough, and irregular atoms, this is enough to explain any difference between the two, however great, even all the difference which is implied between matter and mind. In his strong belief in the unlimited potency of special differences of atomic form and structure to explain the utmost difference between the qualities of various substances, Lucretius reminds us of a capitalist who feels keenly the difference between the commercial value and potency of one bag holding copper farthings and another bag holding golden guineas.

But, after attentively receiving these suggestions, the reader asks—How do the variously-shaped atoms combine so as to produce objects at all? How have they arranged themselves in such marvellous order? After they have united, how is the regularity of their movements kept up? How do colour, scent, and sound come out of the mere juxtaposition of atoms which possess neither colour, scent, nor sound? Finally, how account for the apparition of consciousness out of atoms which are entirely unconscious? To these questions Lucretius attempts no definite answer. In fact, Epicureanism compelled its convert to swallow this dogma without explanation. But how can this fact be accounted for? Simply in this way,—that the scientific mind of Lucretius's day pretty generally accepted atoms as sufficient to prove that the world was not created by God, and that it went on without either guidance or interference by Deity: much as the scientific mind (though far more competent to judge) of the nineteenth century often takes for granted that some other hypothesis, such as Evolution, could

[1] hunc tibi suptilem cum primis ignibus ignem
constituit natura minutis mobilibusque
corporibus, cui nil omnino obsistere possit. vi. 225-38.

science prove it, *must* imply certain important consequences as to morals or religion ;—the connection between the theory and the result to be proved, is overleaped. So the Epicurean argued, if once you allow that *atoms* exist, *ergo*, it must follow that the world made itself.

At the same time, if we consider it closely, we see that Lucretius's theory contains some provisions, not separately put forward, yet constantly implied and often referred to, which seem a little more adequate to the assumed results.

1. Of course Lucretius had no idea of what chemistry has revealed as to the exceedingly complex chemical structure of all forms of organic matter. We know that before a molecule of organic matter can be produced, a most marvellous coincidence of atoms must take place. Lucretius, too, explains the origin of organic matter by a coincidence, though he realizes hardly at all the difficulty implied in it. As we before pointed out,[1] he applies to the combination of the atoms, producing any kind of matter, the word concilium, which is a most unusual term for denoting things without life. (This word must have conveyed to Roman ears the associations of an assemblage of living beings, and thus it implies a rather startling image as applied to dead atoms.[2]) If concilium occurring in dead matter answers to something like chemical combination, on the other hand, when Lucretius speaks of concilium occurring in the production of organic matter, he means it to stand for something like 'spontaneous generation.' Lucretius asserts, as we have seen, that without concilium no kind of matter can be produced. After concilium has taken place, that is to say,

[1] See pp. 42-5.

[2] One might almost suppose that he uses the word, of course only by a metaphor, in its later meaning (which it may have had in common speech even in Lucretius's time), namely, 'generative union.' We cannot press such a line as

Non fieri partum nisi concilio ante coacto, (ii. 935)

occurring as it does in the refutation of an opposite theory.

after molecules have been formed, he supposes these to act partly as a nucleus and assist the formation of other molecules. To exactly the same effect says Professor Clifford, when endeavouring to imagine the production of a molecule of living matter and its development,—'Possibly, however, the molecule of living matter has from the beginning that power, which belongs to other chemical bodies, and certainly to itself, when existing in sensible masses, of assisting the formation of its like.'[1]

2. It is remarkable how often Lucretius applies to the union of the atoms such words as *gignere, genitali concilio,* (continually repeated by him in this sense,) which are a vivid metaphor from living creatures. Similarly he often calls the atoms themselves *genitalia corpora rerum,*[2] 'the particles which beget things.' Again, he constantly applies to the atoms a term never used by his master Epicurus, *semina, semina rerum,* 'seeds,' 'seeds of things,' a name which seems to imply a productive power residing in them. Lucretius must have borrowed this name from Empedocles and Anaxagoras, who applied it to their first principles of matter. The term, though inconsistent with his dogma that matter is 'utterly dead,' is subtly adapted by Lucretius, for beyond question his system implies in the atoms decided tendency and faculties for mutual combination,[3] co-operation, and productiveness.[4] When he describes the work

[1] 'Lectures and Essays,' vol. ii., p. 315.

[2] Cf. i. 632-3, genitalis materies, 'birth-giving matter.'

[3] But when Professor Sellar (p. 319) translates Bk. i. 778-9, as meaning that the atoms, 'in the act of creation, exercise some secret invisible faculty,' he goes far beyond Lucretius's meaning. The words only mean that the atoms are exceedingly small, and have no qualities which sense can comprehend.

[4] Lucretius's explanation of the chance origin and gradual development of the world out of atoms evidently implies some kind of theory of evolution. Perhaps it is best to say that Lucretius realized the difficulties attending such a process only very, very faintly. The following passage from a paper published many years ago, shows to what results the doc-

of the atoms in building up the world and all that is on it, we are often reminded of Tyndall's words, 'The very molecules appear inspired with the desire for union and growth.'

3. Though Lucretius does not specify it separately, it is plain that infinite past time plays a very important part in his

trine of atomic declination may be interpreted: 'An atom that is possessed of volition, and can alter its direction at will, is certainly intelligent; or (since Lucretius flatly denies that atoms are intelligent), it is as good for our purpose as if it were so; it acts as if it were intelligent. When Lucretius assumes that atoms can swerve from their path the least distance possible, it seems a very small thing. Beyond question, if an atom could have free-will, as far as one could imagine, this is just the use it would make of it. It certainly could not use a will of its own in a more modest or less objectionable way. But, in reality, a great deal is granted by this. . . . We now get a truer conception of what the atoms really are. They are not like motes in the sunbeam merely, or the drops in a shower of rain. As we can now imagine them, they are rather like the crowd pouring through the streets of a great city, every individual of which lives and has a will of his own to direct his course, or to turn from the path of the rest as he desires. If we conceive an atom as able to turn to the right or the left at will (and atoms of discretion will, of course, do this on the most necessary and suitable occasions), there is, perhaps, no very great difficulty in their producing the world and its contents. Not more remarkable, perhaps, than for a band of masons and carpenters to build a house. Moreover, only upon this hypothesis in which the atoms become, as it were, tiny workmen, building up the world, can Lucretius's atomic theory be conceived at all possible as an explanation of how the world and all its contains came into being. Like zoophytes building a coral reef, the mechanic atoms ply their mighty toil far beyond our senses' reach. Of course this is absurd, the reader says at once; not merely absurd, but glaringly self-contradictory, for has he not laboured to prove that the atoms are *non-sensile*, only senseless, dead matter? . . . But, perhaps, this whole view of the matter is only an outside one. Perhaps we have not yet grasped Lucretius's real position.'

Atoms such as we have here described, busily building up the fabric of the world, in the fashion of bees toiling over the comb, are little enough like the actual atoms which Epicurus and Lucretius believed in. Yet industrial semi-living atoms like these present a far less reckless inference from Lucretius's doctrine of Declination than does the remarkable theory of 'Spontaneity in things,' which so able a writer as Guyau has inferred from it, and boldly asserts as the actual teaching of Epicurus.

theory. He requires infinite time, along with infinity of space, to enable the atoms 'to try motions and unions of every kind.' His

Ex infinito iam tempore,[1]

answers to Tyndall's 'immeasurable past.' If we allow the existence of infinite space, filled with an infinite number of atoms, which have been in constant motion from all eternity, combining now in one way, now in another, then these atoms must of necessity have tried all possible combinations.[2] At last, therefore, as Lucretius implies, they *must* succeed in striking into the proper motions to produce the world.[3] To Lucretius this appears most natural, *non mirum*, a result which cannot help occurring. When such a combination has resulted, it will maintain itself, and the world thus created will continue to exist.[4] In one passage he says, 'This world has been made

[1] ex infinito vexantur percita plagis. i. 1025.
ex infinito iam tempore percita plagis.
v. 188; repeated at v. 423.

[2] Lucretius brings home to us the endless eternity during which the infinite atomic shuffling has gone on, when he says that in the course of the eternal past, 'the very same atoms of which our bodies are now formed have *often before* been placed in the same order in which they now are.' iii. 854-8.

[3] We may remind our readers of the remarkable passage of Cicero's 'De natura Deorum' (quoted at length on page 79), in which the Epicurean spokesman maintains that an infinite number of atoms in motion amid infinite space is potent to produce the world: 'Cuius (*i.e.* Dei) operam profecto non desideraretis, si immensam et interminatam in omnes partes magnitudinem regionum videretis. . . . In hac igitur immensitate latitudinum, longitudinum, altitudinum infinita vis innumerabilium volitat atomorum.' It is possible that the perusal of Lucretius may have suggested to Cicero the unwonted vividness of these phrases.

[4] namque ita multa modis multis primordia rerum
ex infinito iam tempore percita plagis
ponderibusque suis consuerunt concita ferri
omnimodisque coire atque omnia pertemptare,
quaecumque inter se possent congressa creare,
ut non sit mirum si in talis disposituras

by nature, just as the seeds of things have chanced to clash, entirely of their own accord, after being driven together in many ways, without purpose, without foresight, without result,[1] and at last have filtered through (*colarunt*) such seeds, as, when suddenly thrown together, might become the germs (*exordia*) of great things, of earth, sea, heaven, and of the race of living things.' Most expressive is the phrase *colarunt*, 'have filtered through,' as if the process of infinite atomic combinations must finally separate from the rest and cast together by themselves the atoms fit to create the world.[2]

4. Lucretius attempts circumstantially to describe the origin of the world. He does not leave the imagination altogether without assistance in its efforts to realize how this came about.

deciderunt quoque et in talis venere meatus,
qualibus haec rerum geritur summa novando. v. 187-94.

tandem deveniunt in talis dispositura,
qualibus haec rerum consistit summa creata,
et multos etiam magnos servata per annos,
ut semel in motus coniectast convenientis. i. 1027-30.

[1] *Temere, incassum, frustra* (of which we quote Mr. Munro's expressive rendering) occurring after 'ipsa sponte sua, forte offensando.' How emphatically, by this reiteration, does Lucretius assign the world's origin solely to the play of chance. ii. 1058 ff.

[2] We may point out (what we have not seen noticed elsewhere) that this very expressive metaphor of an atomic sifting is borrowed from Democritus. According to Democritus, it is a universal law that like atoms are attracted to each other, *τὰ ὅμοια πρὸς τὰ ὅμοια*. The cause of this is a merely mechanical one. The eternal whirling motion of the atoms has a tendency to bring together those which are like each other in size and form, and, as it were, to sift the atoms into separate heaps. In a fragment given in Sextus Math. (vii. 116 *sq.*), Democritus himself uses this very metaphor of an atomic sifting. The atoms, he says, run like to like, 'as you may see either in the case of seeds which are being winnowed in a sieve, or in the case of pebbles on the sea-shore; for on account of the whirling of the sieve, beans are separated and go with beans, barley with barley, and wheat with wheat; and on account of the motion of the waves, the longish pebbles are driven to the same spot as the longish ones, and the round with the round.'

His explanation reminds us not a little of the nebular theory now in fashion. According to this, a part of the primeval nebula, or fire-mist, which filled the whole space now occupied by our solar system, at some time broke off from the mass, and began to revolve separately. This was the beginning of our world. Gradually this portion of nebula condensed and grew cooler, till, in the course of millions of years, it became fit to be the dwelling-place of animals and men. Whether accurate or not, Lucretius's account of the matter is not much less coherent.

After the first-beginnings had tried all kinds of combinations, 'Those particles at length meet, which, being suddenly carried together, often become the rudiments of great things, of earth, sea, and heaven, and of the race of living creatures.' At the first stage of the world's history, neither sun, star, earth, air, nor sea was yet in existence. 'Nor could any substance be seen like to our substances, but a kind of strange storm and medley made of atoms of every kind, whose lack of harmony caused a conflict, and disordered their interspaces, passages, connexions, weights, blows, clashings, motions, because, by reason of their unlike forms and different shapes, they could not all remain thus joined together nor unite in harmonious motions.' Gradually, however, this chaos ceased; like atoms joined like, and thus earth, sea, air, and ether began to exist separately. How did this come about? Lucretius explains the matter most ingeniously. 'First of all the several bodies of earth, because they were heavy and closely entangled, met together in the middle, and took up all the lowest positions, and the more they grew entangled, and came closer together, the more did they squeeze out those particles which were to make the sea, stars, sun, and moon, and the walls of the great world, for all these are formed of smooth and round seeds, and of far smaller[1] elements (*i.e.* atoms) than is the earth.' Last of all,

[1] We may compare the description of the soul-atoms. See chap. vi.

the fiery ether burst out of the pores of the earth, just as we now see mists rise from earth in the morning; and as the mists unite and vault the sky with clouds, so these particles of ether, rising because of their lightness, united and formed a vast shell, fretted with golden fires, surrounding the earth in a wide circuit. 'In this way the light and nimble ether grew into a body, and, encircling the rest, arched itself on every side, and spreading widely in all directions round about, it thus shut in all the rest with its greedy embrace.' After ether, rose the elements of sun and moon; these, being too light for the earth, and 'not light enough to glide along the upmost borders' and join the ether, join in spheres, and float between the two. The departure of all these particles from the earth caused a hollow, and into this the waters collected and formed the sea. 'And day by day, the more that the heats of ether round about and the rays of the sun compressed the earth on all sides by their repeated blows into a close mass—so that, being thus buffeted, it was condensed and consolidated at its centre—ever the more did the salt sweat, forced out from its body, by its oozing, increase the sea and its floating fields, and ever the more did those many particles of heat and air escape and fly abroad, and, by condensing, form the high and shining quarters of the heaven far away from earth.'[1]

The earth, thus formed, rests stationary in the middle of the world. How, then, is it kept in its place if it be surrounded below, as well as above, by air and ether, both lighter than itself? Lucretius explains this in a most ingenious way.[2] Earth, air, and ether, he says, are not bodies alien to each other, and forced together. On the other hand, 'the earth was conceived together with the air at the first birth of the world, and is a fixed portion of that world, just as our limbs are seen to be to us.' Earth, air, and ether are parts of one organic whole. 'They cohere together with common roots, conjoined and

[1] v. 432-508. [2] v. 534-63.

formed into one being from the beginning of their existence.' The earth 'is embodied, and lives' in the air and ether.[1] In this way the earth is able to rest in the middle of the world, because its weight gradually passes away and is lessened, from the centre outwards. Thus the earth does not weigh down the air beneath it, any more than our limbs or head burden our body. The air and ether, though so subtle in texture compared with the earth, are able to hold it up because the earth and 'the airy portions of the world' are closely conjoined and form one being—indeed one organism. Similarly the soul, though extremely light compared with the body (*tenuissima vis animai*), is able to support, and even at will to lift it with a sudden leap, because it is so closely united with it,

propterea quia tam coniuncta atque uniter apta est.

Thus it comes that the earth floats in the middle of a sea of air surrounded by an outer space of ether.

How then, we cannot help inquiring, does the world hold its place in the universe? As we have before shown, it is difficult to be certain how Lucretius explained this. It seems to us almost certain that he supposed the outer globe of air and ether to buoy up the heavy core of earth by their extreme lightness, and thus to enable it to float amid the ocean of atoms. No doubt all things tend downwards from gravity, he has said, but the blows of the atoms streaming up from below to feed the world with matter would be more than sufficient to keep so buoyant a sphere suspended in its place.

It is thus, as we have seen, to a chance combination of the atoms that Lucretius supposes our world to owe its birth. And, in the same way, new worlds are being ever struck into being

[1] We quote Mr. Munro's rendering. Lucretius speaks of the earth as

ex ineunte aevo coniunctam atque uniter aptam
partibus aeriis mundi quibus *insita vivit*. v. 537-8.

Lucretius certainly conceives the world (not the earth) as an organism *of a kind*.

by the atoms ever battling in infinite space—bubbles formed on the surface of the infinite torrent of matter, and lasting for a while ere they break. Lucretius does not suppose that all the worlds are like our own earth. 'In other parts of space there are other worlds and different races of men and kinds of wild beasts.'[1] 'Throughout the universe the various worlds are formed on various plans.'[2] Leibnitz held the remarkable opinion (which has been designated by the term Optimism) that ours is the best of all possible worlds. Lucretius does not decide this point.

What, then, does Lucretius conceive our own world—the little corner of infinite space and infinite matter wherein we live—to be like? Every world (that is to say, every system containing an earth, heaven, and heavenly bodies like our own, all which he and Epicurus include in the name 'world')—every world is walled in by an outer circuit of ether, studded by stars, which Lucretius conceives to be fires fed by the substance of the ether. This outer rampart of ether, set with the star-fires which it feeds, Lucretius is fond of calling by a noble phrase, 'The flaming walls of the world,'

Flammantia moenia mundi,

and he conceives it to act as a barrier against danger to the world from without. In his description of the end of the world, it is the *moenia mundi* which first 'are stormed' and give way.[3] The stars move either in the ether,[4] or along with it. Within this is an inner circle of air, in which move the sun and moon,

[1] necesse est confiteare,
esse alios aliis terrarum in partibus orbis
et varias hominum gentis et saecla ferarum. ii. 1074-6.

[2] Per omne
in variis mundis varia ratione creatis. v. 527-8.

[3] i. 1102-3; ii. 1148-9.

[4] sive ipsi serpere possunt
quo cuiusque cibus vocatatque invitat euntis,
flammea per coelum pascentis corpora passim. v. 522-5.

floating midway between earth and ether. The earth itself occupies the centre; and since air, ether, and heavenly bodies are all formed from the lighter or more fiery particles which have escaped from the earth, probably Lucretius supposes the earth to form a very large part of the world, and to occupy most of the space contained within it.[1] This whole system—earth, air with sun and moon, and the outer shell of ether with its stars—Lucretius calls the world.

Though the all-enclosing ether 'holds the world in its greedy embrace,' yet it allows free passage to the atoms which, as we have seen, are ever streaming up out of the infinite void to feed the world and keep it in being. 'On every side through all the pores of the ether, and, as it were, through the breathing places of the great world all round it, a free passage out and in is given to the atoms.'[2] Thus is our world vitally related to

[1] Lucretius does not in the least realize what the actual size of the sun and heavenly bodies is.

> nec nimio solis maior rota nec minor ardor
> esse potest, nostris quam sensibus esse videtur.

The stars also are about the same size as they appear to us, perhaps a very little larger, or a very little less.

> scire licet perquam pauxillo posse minores
> esse vel exigua maioris parte brevique.

This sounds ludicrous to us. At the same time, if we examine the facts of observation quoted by Lucretius in support of this conclusion (namely, experiments as to the apparent size and brightness of fires on the earth when observed from a distance), and the way in which he treats these facts, we see that his method of approaching this question (though it has led him to such a conclusion) is by no means so very unscientific as we might suppose. See v. 564-91.

[2]
> undique quandoquidem per caulas aetheris omnis
> et quasi per magni circum spiracula mundi
> exitus introitusque elementis redditus extat. vi. 492-4.

Compare the following lines referring to the human body—

> donec materies, omnis concussa per artus,
> vitalis animae nodos a corpore solvit
> dispersamque foras per caulas eiecit omnis. ii. 949-51.

the infinite outer spaces surrounding it. It must be ever fed out of the infinite surrounding sea of matter, or it must waste away, break up, and die.

One cannot help realizing how different is the picture of the world, as painted by Lucretius, from that which Epicurus would have drawn for us, hard, cold, and colourless. In all scientific details, Lucretius follows his master step for step; but how different is the result! Lucretius fills us with a sense of the wondrous origin and maintenance of the world, of its hazardous tenure of existence, and of the vast encircling forces whose counterpoise preserves it, but which may at any moment shatter it amid their play as it were, of the piteous weakness of human life in the lap of powers so mighty and so reckless, of the beauty of all natural landscape, of the richness and loveliness of the world which, though chance-born and existing amid such dangers, besieged with hostile forces on every hand, flowers into fresh beauty and life every spring-time, though another year may never come to it. Nothing of this could Epicurus have given us. If we look merely at a single point of the Epicurean cosmogony, the notion of the ether which Epicurus adopted from the early Greek philosophers,—how majestic a conception is it as Lucretius is fond of picturing it, 'The flaming walls of the world,' a fiery bastion guarding the earth, and walling it in from the infinite outer seas of space, filled with infinite atom-dust. The imagination must be dull indeed that is not touched by its grandeur. Before us lies a much-worn copy of an early edition of Lucretius's poem, on the yellow fly-leaf of which some lover of the poet, centuries ago, has written in antique letters the words—

Extra flammantia moenia mundi!

5. Lucretius finds little difficulty in giving an atomic explanation of the origin of life upon the earth. He has shown already that life is only a mode of motion of dead matter. He

asks—seemingly without any shadow of misgiving—Do we not see life produced on a small scale before our eyes at any time, when rotting substances under sun and rain bring forth worms? In the same way, the earth long ago produced all the living creatures. It is only because the earth is old, and past her season of bearing, that she now produces only small creeping things. Long ago, 'In the fresh time of the earth and ether,' she was able to bring forth mightier offspring, even 'the huge bodies of wild beasts.' It was then, in her lusty youth, that she gave birth to all creatures that enjoy the light. The young earth first put forth grass and trees, and then it produced higher forms of life, living creatures of every kind, which sprung up, 'many in number in many ways, after diverse fashions,'

multa modis multis varia ratione coorta.

First the earth produced eggs from which birds were hatched in the spring-time.[1] Men came into existence in a more complex fashion. 'For much heat and moisture then abounded in the fields, and therefore, wherever a suitable spot occurred, there would grow wombs fastened to the earth by roots.' Within these wombs were infants, and as the natural warmth of the living creatures in course of time caused these *uteri* to open, then just in the same way as happens to the mother when a child is born, 'nature would turn to that spot the pores of the earth, and compel it (the earth) to yield from its open veins a liquid exactly similar to milk.'[2] In this remarkable

[1] After l. 804 there is apparently a lacuna. The account of the *modi multi* in which various animals were produced by the earth, which Lucretius begins to explain with a *principio* at l. 801, particularizing farther by *tempore verno* at l. 802, is plainly incomplete. Apparently, too, from the general context, as well as from l. 801, Lucretius must have meant to tell us something about the *order* in which the different creatures came into existence.

[2] multus enim calor atque umor superabat in arvis.
hoc ubi quaeque loci regio opportuna dabatur,

notion of *uteri* growing on the earth, is there any reminiscence of the Empedoclean doctrine that the different bodily organs came into existence separately and apart from each other?[1] It has recently been pointed out[2] that the notion of streams of milk flowing from the earth to feed her infant human offspring may have been suggested by the myths of the Golden Age. If this be the case, it may have seemed not so unnatural to Lucretius's contemporaries, who were familiar from their childhood with the legend of a far-off happy age, when—

flumina iam lactis, iam flumina nectaris ibant.

But Epicurus probably derived the notion from Archelaus,[3] who taught that 'living creatures were produced from the earth when it was hot, and that the earth sent forth a thick mud resembling milk for their food, and that she produced men also in this way.'[4]

We must not estimate by the present the powers of production which the earth possessed in its lusty youth. It was then fuller in substance, and contained 'many seeds of things' which it has now lost.

crescebant uteri terram radicibus apti;
quos ubi tempore maturo patefecerat aestus
infantum fugiens umorem aurasque petessens,
convertebat ibi natura foramina terrae
et sucum venis cogebat fundere apertis
consimilem lactis, sicut nunc femina quaeque
cum peperit, dulci repletur lacte quod omnis
impetus in mammas convertitur ille alimenti. v. 806-15.

[1] Aristotle, 'De Coelo.,' iii. 2.

[2] By Mr. Benn, in 'Westminster Review,' April, 1880, p. 340.

[3] Diogenes tells us that, according to Diocles, 'of all the older philosophers Epicurus was most in agreement with Anaxagoras (although in some points opposed to him) and with Archelaus, the teacher of Socrates' (Diog. L., x. 12).

[4] *γεννᾶσθαι δὲ φησι τὰ ζῶα ἐκ θερμῆς τῆς γῆς καὶ ἰλὺν παραπλησίαν γάλακτι διον τροφὴν ἀνιείσης· οὕτω δὲ καὶ τοὺς ἀνθρώπους ποιῆσαι* (Diog. L., ii. 17). This passage is referred to by Reisacker ('Questiones Lucretianae,' 1847, p. 64).

multa fuere in terris semina rerum
tempore quo primum tellus animalia fudit.[1]

About the same period, the earth produced both animals and men,[2] and after this, 'like a woman exhausted by age, she ceased to bear.'

In this era of productiveness the earth 'attempted to produce' (*creare conatast*) all kinds of monstrous creatures, some half-man, half-woman, some without feet or hands or mouth or eyes; some with their limbs adhering to the trunk. All these the earth produced, *nequiquam*, says Lucretius, 'all in vain,' since they were unable to preserve themselves from danger, or to secure food, while some were isolated productions, and others organically unfit to reproduce themselves. Lucretius appears to think that every kind of monstrous creature which could possibly exist, must have been produced by the earth at this time, while out of them all, only the few adaptations were preserved. Just as the atoms tried every kind of combination possible before they could strike into the order necessary to produce the world—

namque ita multa modis multis primordia rerum
omnimodisque coire atque omnia pertemptare,
quaecumque inter se possent congressa coire—

so living creatures, misshapen with every degree of bodily deformity possible within any given type, must have been

[1] v. 916-7.

[2] genus ipsa creavit
humanum atque animal *prope certo tempore* fudit
omne quod in magnis bacchatur montibu' passim
aeriasque *simul* volucres variantibu' formis.
sed quia finem aliquam pariendi debet habere,
destitit, ut mulier spatio defessa vetusto. v. 822-7.

Do the words prope certo tempore refer to the production of animals only, as their position would imply, or to that of man also? Here, as in the preceding section (v. 783-820), which is probably unfinished, Lucretius speaks vaguely on this point.

produced before the few organic forms fit for continuance could be selected by the test of adaptation. 'Every other monster and portent of this kind earth would produce, but all in vain,'—

> cetera de genere hoc monstra ac portenta creabat,
> nequiquam,——

A new vista this in the long process of Evolution, and not a lovely one to look down,—one that haunts the mind with the possibility of infinite aimless suffering on the part of these helpless and ludicrous monsters created, *nequiquam*, that is to say, created only to be tortured out of existence by *Natura* as soon as possible.[1] Darwin's theory of Evolution is at least not distorted by any such horrible fancy as this. The notion is hideous, and, were it true, even the survivors fortunate enough to be left by such a process could hardly congratulate themselves. Such waste of suffering might even make us sympathize so far with Schopenhauer's declaration, that consciousness is the great mistake of the universe. The earth is a grim enough place, but the fancies of certain philosophers make its shadowy corners even darker. For our own part we find 'l'homme qui rit'—the monster who was made by the devilishness of man and not of nature—to be sufficiently horrible.—Lucretius was not unfeeling, but the reverse, and if he only half realized the horrors of a long era when the earth was covered with millions of crippled and misshapen creatures, ever fatally baffled by their own deformity and helplessness, created only to suffer, it would be no wonder if his fancy had been visited by processions of sad and misshapen figures, a bestial or human Goblindom, horrible as the *mirae figurae*, the 'strange shapes,' which, as he complains, haunted and made terrible his dreams.

Lucretius insists that all this production of organic forms

[1] Cf. v. 877,—

> donec ad interitum genus id natura redegit.

must have gone on according to Law. Among all the abortive types which Nature may have produced, no such thing as a Centaur, half-man and half-horse, or the Chimæra, in part a lion, in part a dragon, in part a goat, can ever have existed. It is not possible for the limbs of different creatures to be combined in a single frame, because 'each thing goes on after its own fashion, and all preserve their distinctions after the fixed law of nature.' [1]

The monsters of necessity perished. But many kinds of animals must also at one time have lived, which in the fierce struggle for existence must have gone to the wall. Only the possession of some special quality, such as courage, speed, craft, has enabled each race to exist and continue its kind. Some animals, such as dogs, oxen, sheep, are useful to man, and have in return been preserved through his protection. Such as neither had natural properties to depend on nor were useful to man, 'lay exposed as a prey for the gain of others,' and died out in the fight for existence.

Donec ad interitum genus id natura redegit.

'That species Nature utterly destroyed.' Lucretius sees as clearly as any Darwinist the pitilessness of Nature. In attributing the dying out of such species, not to internal unfitness for life, but to outward conditions, the competition with other individuals or species in the struggle for food, there is certainly more than a glimpse of Darwin's great principle.

When we review Lucretius's explanation of the origin and history of life upon the earth, we see that it is based on a clear perception of Darwin's doctrine, that in the organic world none but adaptations continue to exist, because these alone have been able to perpetuate themselves, while those types which show more or less imperfect adaptation have perished alto-

[1] sed res quaeque suo ritu procedit et omnes
foedere naturae certo discrimina servant. v. 923-4.

gether. Beyond question, Lucretius had a firm grasp of this other central doctrine of Darwinism.

Are we then to give Epicurus the entire credit for this anticipation of modern science? Certainly we cannot. There is no doubt that Epicurus, in his exceedingly shrewd eclecticism, which borrowed, among other things, the notion of ἀποῤῥοιαι, or 'effluxes,' from Empedocles to explain the phenomena of perception, making it the basis of his psychology, and the notion of atoms from so opposite a thinker as Democritus to make it the basis of his science, in each case no doubt correcting and greatly developing the opinion of his masters,—there is no doubt that Epicurus adopted his doctrine of the origin of life and species from Empedocles, who believed in the production of organisms immediately through the fortuitous combinations of the elements,[1] and in the final preservation of the fittest, while the others perish as they are produced. But when we look into the matter, we find how crude was Empedocles's notion compared with Epicurus's development of it. According to Empedocles the several organs of animals, such as eyes or arms, came into existence separately, each without bodies, and wandered about thus, till the combining tendency of nature united them in complete organisms. Empedocles believed also in the past existence of unnatural monsters, with limbs heterogeneously blended, whose existence Lucretius shows to be against law and impossible. Both of these notions Epicurus sets aside. Beyond question, ancient Greece can not unfairly claim to have originated and developed the doctrine which Darwin has revived under the name of Natural Selection,' and which now exerts so mighty an influence.

[1] At the same time, in Empedocles's hands, this doctrine is not connected with any disbelief in Divine power. It was Empedocles whom, from his intensely religious mental attitude, Aristotle calls a scientific θεολόγος. Moreover, we must remember that Empedocles's material elements, which he calls Daemons, are Divine.

6. Lucretius, as we know, does not believe in Divine action. Can it be said that he had any notion of Evolution? It is certain that his philosophy implies Evolution; and Lucretius may fairly be taken to support it, though of course his conception of such a process must have been of the vaguest kind, speaking from a modern stand-point. It is easy to see that some minds in antiquity—members of the Epicurean sect in particular—felt a strong repugnance to the popular belief of a Divine Being constructing the world after the methods of a human artificer. The Epicurean spokesman in Cicero's treatise, 'On the Nature of the Gods,' expresses this thought very distinctly.[1] In very remarkable language he says, 'The philosopher who taught us all we know' (that is, Epicurus) 'has shown us . . . that the world was made by nature, and that there was no need of any workshop to construct it in;' and again, 'In this infinitude of breadth, length, and height are fluttering about an infinite number of innumerable atoms, which, though with void space interposed between them, still cohere, and, clinging to one another, are massed together, whence are made those forms and shapes of existing things, *which you think cannot be produced without the help of bellows and anvils.*' Could Tyndall even have asserted more pointedly the self-developing, formative tendency in Matter, which dispenses with an outside Divine mechanist? Lucretius, too, casts teleology away. Neither the world nor the human body, he says, show any trace of design;

[1] See especially chaps. viii. and xx. of Bk. i. The original of the passage quoted is most expressive in its wording. 'Docuit enim nos idem qui cetera, natura effectum esse mundum: nihil opus fuisse fabrica. . . . In hac igitur immensitate latitudinum, longitudinum, altitudinum infinita vis innumerabilium volitat atomorum, quae interiecto inani cohaerescunt tamen inter se et aliae alias apprehendentes continuantur, ex quo efficiuntur eae rerum formae et imagines quae vos effici posse sine follibus et incudibus non putatis' (i. c. 53-4). According to Lecky, Cicero's account of the Epicurean system is one of the grandest examples to be found of 'sublime and scrupulous justice to opponents.' ('History of European Morals,' vol. i., p. 185).

our eyes, feet, hands, were not made for our sake that we might see, walk, labour. Nothing has been made for the use of men.'[1] Commenting on the fact that men, animals of every kind, grains of corn, shells on the sea-shore, if we compare specimens of them together, are all different one from the other, he points out that Nature's style of production differs from that of man. All these objects, he says, 'Exist by nature, and are not manufactured by hand after the exact model of one.' Lucretius, like Tyndall, is opposed to the conception of an 'Artificer in the universe, fashioned after the human model, and acting by broken efforts, as man is seen to act.' Had Lucretius become acquainted with the modern theory of Evolution, beyond question he would have eagerly embraced it, as filling the gap in his system, only that he seems, as we have suggested, not to have been conscious that there was any gap. His theory of atoms, and his principle that 'Nature is seen to do all things spontaneously of herself, without the meddling of the gods,' clearly point this way. They show us that a wholly new conception of Nature, as the one and self-working power, manifesting itself in the world, was abroad in the air. Such a conception, even though Lucretius cuts it clear from the notion of one all-ruling and ever-working Divine Will, could not fail to work in men's minds as soon as they took it in. It must speedily have rooted itself in the thought of every intelligent reader of his poem, and have grown up, and born fruit. Lucretius himself may formally insist that his *Natura* is only an abstraction, yet it is an idea which perceptibly modifies and influences his view of the world at every moment.

In reviewing Lucretius's theory of the atomic origin of the world and the life upon it, and on comparing it with modern theories, two things strike us.

In the first place, Lucretius explains the origin and continuance of life in the same way as Clifford explains the origin of

[1] See Bk. iv. 823-57, and Bk. v. 156-94.

the first molecule of living matter. Clifford's words may hold true for both, when he says that the first molecule of protoplasm was 'produced by a coincidence and preserved by natural selection.'

Secondly, would Lucretius ever have approved of any system of Evolution which involves the gradual transformation of species into each other? He has stated repeatedly and most strongly the permanence and distinctness of all natural types. His definition of species certainly does not include the capacity of varying.[1] Apparently he believes that a living creature must cease to exist on the slightest change of its form or type. For both organic and inorganic things, his conception of Law made him rigidly realize the consequences of *change*, which he has thus expressed: 'Whenever a thing is changed, and leaves its limits, this brings at once the death of that which was before.'[2] Within these limits Lucretius clearly realizes that there is free play for a certain amount of variety; no two men, beasts, birds, fishes, grains of corn, shells on the sea-shore are exactly like each other—

> quorum unum quidvis generatim sumere perges
> invenies tamen inter se differre figuris.[3]

[1] Speaking of the Darwinian principle of an innate tendency to variation in animals, Mallock has well remarked, 'It will thus be seen that the Darwinian theory is an advance on, and differs from the Lucretian mainly and essentially in this—the way in which the variety is produced, which is the subject of the selecting process common to both systems' (Lucretius, p. 50). See also Lange's remarks on Empedocles (vol. i., chap. i.).

[2] nam quodcumquae suis mutatum finibus exit,
continuo hoc mors est illius quod fuit ante.
ii. 753-4; repeated at i. 792-3.

[3] ii. 333-80, where Lucretius dwells on the distinct individuality of all living things, and in illustration of this draws a most pathetic picture of the cow seeking the calf which has been taken from her, and whom no other calf will satisfy.

At the same time he holds that no species can by any possibility merge into another, unless the individuals belonging to it be first broken up into their component atoms, and then refashioned by the moulding hands of Nature after a different model. Had the theory of variation of species been actually presented to Lucretius by any contemporary philosopher, we can easily imagine him denouncing it from the stand-point of Epicurean science, because it contradicts the law of fixity of type, criticising it shrewdly enough, and asking where in Nature we see one species passing by long leaps, or even by short ones, into another;[1] *Perdelirum esse videtur!* would have been his very mildest sentence.

As we have already shown,[2] Lucretius connects the permanence of species with the absolute changelessness of the atoms. In the constant reproduction of all the distinctive features of every kind of living things, in one generation after another, he sees a proof that the atoms are indestructible. Reasoning on the very same principle as Lucretius, Clerk-Maxwell has said, 'No theory of evolution can be found to account for the similarity of the molecules, for evolution necessarily implies continuous change, and the molecule is incapable of growth or decay, of generation or destruction.'

One can hardly think about Lucretius and his belief in the purely atomic origin of the world without remembering how similar a position is held by some others at present. Their line

[1] How entirely the evidence is still lacking for any sign of the stepping-stones which should lead from one organic form to another—and higher, no Evolutionist can deny. Mr. W. K. Parker has stated the real difficulty admirably as follows:—'We may liken the present animal creation to a high staircase; monad the first step, Homo the top one. Homo, however, does not—nor, as for that, do the others—appear to have gone up obliquely, but vertically; as if the staircase had been constructed perpendicularly, and all the substructure taken away afterwards, for it is all gone' (quoted in Mr. Reynolds's work, 'The Mystery of Miracles,' p. 148).

[2] See p. 20.

of argument stretches boldly in the same direction, while it is broken by fewer gaps. In the following chapter we shall try to give a notion of the modern argument for Materialism, where it labours to build in the gaps which Lucretius was compelled to leave.

CHAPTER V.

HOW MODERN SCIENCE BRIDGES OVER THE GULF BETWEEN ATOMS AND LIVING THINGS. THE CONTROVERSY AS TO THE POTENCY OF MATTER.

AS we have seen, Lucretius, in spite of all his ingenuity, is compelled to leave an impassable gulf between dead atoms and the world with all its life. The modern Materialist has borrowed from science abundant store of materials with which to build a bridge over the gulf, and has spared no labour in the attempt to construct it. Nothing will help us to realize this subject more vividly than the famous Presidential address [1] in which Professor Tyndall has set forth in full the arguments by which modern Materialism defends its position. These are, at the same time, the arguments of the Evolutionist, and it is in that light that Tyndall presents them. A more powerful exposition of what these reasons are we can hardly conceive. We do not require to decide whether he is himself a Materialist, but the line of reasoning which he has stated with so much force is that followed by most who are such. To some Evolution is but an argument for Atheism. In the course of his long address, Tyndall endeavours to show that the old conception of an outside Demiourgos, a Divine workman conceived by man in his own likeness, shaping and fashioning the world, has passed away before our increased knowledge of nature. Formerly naturalists, he says, believed that a special creative act was necessary to account for the appearance of each new group of organisms.

[1] Delivered before the British Association at Belfast. (Longmans and Co., 1874.)

But to the natural philosopher, who has no prejudices of 'previous education,' no contrivance like that of a human artificer is to be seen in the productions of nature. The method of Nature is not like that of man. She has her own method,—it is that of Evolution, a constant unfolding.[1] For this purpose no Demiourgos need be present; we may 'detach the Creator from his universe.'[2] (We may remind our readers how Epicurus

[1] The method of Nature as an evolution from within is not more keenly realized by Tyndall than it was by Bruno, centuries ago, and without the light of modern science. 'I affirm,' says Bruno, 'the physical universal efficient to be the universal Intellect which is the primary and principal faculty of the Soul of the world; that Soul being, on the other hand, the universal form of this Intellect.' . . . 'This is that One which fills the great whole, which illuminates the universe, which directs Nature to produce its species in the way which is most suitable. So that it has the same relation to the production of *natural* things as our intellect has to the corresponding production of rational species. That which Plato . . . and Plotinus called the Fabricator of the world, . . . the Father or Progenitor, ought in reality to be called the Internal Artificer, seeing it forms the matter and the figure from within: from within the seed or the root it gives forth or unfolds the stem; from within the stem it forces out the boughs; from within the boughs it forces out the branches; from within these it pushes out the buds; from within it forms, shapes, and interlaces, as with nerves, the leaves, the flowers, the fruits; and from within, at appointed times, it recalls its moisture from the leaves and fruits to the branches, from the branches to the boughs, from the boughs to the stem, from the stem to the root. And there is a like method in the production of animals.' (From Bruno's treatise, 'Della Causa Principio ed Uno,' 'On the One Sole Cause of Things,' translated in Plumptre's 'History of Pantheism,' vol. i., p. 358.) Compare also Bk. viii., c. 10, of Bruno's Latin poem 'De Universo et Mundis.'

[2] In this assertion Tyndall goes beyond Darwin, but the charge has been brought against the 'Origin of Species,' that while Darwin places a Creator at the beginning of the process of organic evolution, He is there introduced 'for ornament' or rather 'to conciliate theological prejudices' than because He has any real work to do. He 'sits passively by and watches laws which execute themselves,' and, in short, 'He is a monarch that reigns but does not govern.' On the Darwinian hypothesis animals and men are in reality produced according to mere Chance.—But Darwin's theory of Evolution does not necessarily dispense with a God. Darwin has written 'It may be metaphorically said that natural selection is daily

'detached' his gods from the world altogether.) The marvellous adaptations, the so-called proofs of design, are but the offspring of Nature, after her own method bringing forth all things of herself. But how can this take place? When we wish to account for the world and the life upon it, two courses, and two only, are open to us. 'Either let us open our doors freely to the conception of creative acts, or, abandoning them, let us radically change our notions of matter.'[1] After putting the alternative thus, Professor Tyndall enthusiastically expresses his belief in the latter conception. 'Abandoning all disguise,' he says, 'the confession that I feel bound to make before you is that I prolong the vision backwards across the boundary of the experimental evidence, and *discern in that matter,* which we in our ignorance, and notwithstanding our professed reverence for its Creator, have hitherto covered with opprobrium, *the promise and potency of every form and quality of life.*' The difference between the views of Tyndall and Darwin on the subject may be stated thus: Darwin conceives the world at the beginning as full of dead matter, and in the midst of this, one primordial living form, capable of self-development into other living forms. Tyndall prefers to conceive at the beginning a world of matter in which there is no need of a

and hourly scrutinizing throughout the world every variation, even the slightest; rejecting that which is bad, preserving and adding up that which is good, silently and incessantly working whenever and wherever opportunity offers at the improvement of every organic being.' We may ask, Is such a method of Creation necessarily inconsistent with the never-resting, incessant watchfulness of Mind, for whose care nothing is too small, whose power and wisdom nothing taxes too much? If Evolution causes 'one or a few primordial forms' to develop by a process of infinitesimal changes into all the rich variety of the world's organic life, could such a method ever grow up out of the reckless play of Chance?

[1] In passing we must mention that this alternative is unfair. The fallacy seems to lie in the word 'freely,' by which Tyndall would force upon us the most vulgar and unscientific of the existing conceptions of creation. Of course his answer is that the creation of a few forms involves anthropomorphism as much as the creation of a multitude.

single living germ, for there is no dead matter; all matter is is living and able to evolve from itself living forms of every kind. Thus Tyndall rejects Darwin's hypothesis of the creation of a few living forms. If matter is animated, then such forms are not needed. But by this Tyndall only throws creation further back—back to matter which must have been endowed with such powers of producing life. Whence, then, did matter receive such powers? But this question he will not answer. Only after he has returned an answer can we decide whether his position is necessarily inconsistent with Theism.[1] Up to this point it certainly is not. Of course, if matter is not created, and as Professor Tyndall also implies, a God exists, it does not seem possible to evade the conclusion that matter is eternal, and God identical with matter. Perhaps Tyndall may be not unwilling to be ranked in the same class to which he has himself assigned Bruno, namely, as a 'Pantheist.' Yet we must not forget that there is a 'higher Pantheism' as well as a lower. The opinions expressed in the address are not inconsistent with the existence of a Creator. Yet a deity such as Tyndall conceives, who is, in a very important sense, wholly detached from his universe, who cannot hear the voice of prayer, and whose worshippers must neither 'seek nor expect' aid (Professor Tyndall prefers to call it 'personal profit') in the hour of need, does assuredly remind us of the ignoble Epicurean gods 'who lie beside their nectar' and take no heed of men.

How, then, does Tyndall take the step from the moving molecules to the existing world with its beautiful and complex living forms? Accepting the doctrine of atoms (much as Lucretius held it) as the basis for constructing the universe, how does he explain the process? We can only refer to the

[1] On this subject we may refer, without expressing any opinion, to the Appendix to Picton's 'New Theories and the Old Faith,' which contains a very remarkable note 'On the Development Theory in relation to the Soul and Immortality.'

three most important or most striking points in his argument. The first question that the Materialist is called upon to answer is, How can matter produce and account for thought and consciousness? He finds little difficulty here,—Life, he says, cannot be conceived of apart from matter. Divorced from matter where is life to be found? Vibrations in the brain invariably accompany thought, and actually are thought. He gives many an argument on the materialistic side to prove that there is no other self different from the brain-self, that the brain is the man. He can ask several questions very difficult to answer. Can you form a mental picture of any of the percipient powers, apart from the organism through which it is supposed to act? If consciousness is a proof of the true self being distinct from the body, what do you say of the whole body being deprived for a short time of consciousness, as in the case of fainting? If a change of brain makes an exemplary man a murderer, is it possible that the true self can remain as before, and that with the physical change, his character is not altered? The brain cannot be viewed as a mere instrument, like an eye-glass or a staff. It is more. The union between brain and soul is so close that the conditions of the body react on the soul. Therefore, says the Materialist, we are but

'Only cunning casts in clay.'

(Professor Tyndall has with considerable force argued the opposite side as well—that molecular processes can never wholly account for consciousness.) Darwin and Herbert Spencer have sought to show how the processes of Evolution and of gradual development from lower to more perfect organisms fill up the gulf between the monad and the man, with senses, intellect, and consciousness complete.

But Tyndall has stated a second argument with great ingenuity in his endeavour to show by what scientific reasoning the Materialist attempts to account for the origination of life, a point which, as he says, has been but 'lightly touched upon, if at all, by Mr. Darwin or Mr. Spencer.' He takes a very striking illustration. If we break a magnet, he says, we find in each fragment two poles. If we continue the process of breaking, we find that each part, however small, carries with it, though in a smaller degree, the polarity of the whole. When we can break no longer, we prolong the intellectual vision to the polar molecules. 'Are we not urged,' he says, 'to do something similar in the case of life?' The farther back we trace the line of life, we find it approaching nearer and nearer to what we call the purely physical condition; that is to say, we reach organisms of the very simplest type, like the *protogenes* of Haeckel, in which, so far as we can discern, 'the vital action is almost wholly physical.' But after we have thus reached the very simplest known organism, Tyndall bids us cross the border land of sense and prolong the intellectual vision from the more perfect organisms to the very lowest ones in which life can be conceived to originate. Scientific men can justify scientifically their belief in the potency of matter, under the proper conditions to produce organisms. But they will frankly admit that they cannot point to 'any satisfactory experimental proof that life can be developed save from demonstrable antecedent life.' Bastian's experiments[1] show that spontaneous generation does take place in the sense of life being produced from dead organic matter,—what he calls Archebiosis. But no one has shown that spontaneous generation takes place in the sense of Heterogenesis, that is, the production of life from what we call inorganic matter. But, says Tyndall, scientific men, as already indicated, draw the line from the highest organisms through lower ones down to the lowest, and it is the prolongation of the line by the intellect beyond the range of the senses that leads them to the conclusion which Bruno so boldly enunciated, viz., that matter can originate

[1] See 'Contemporary Review' for September, 1874.

life. 'Believing, as I do, in the continuity of Nature, I cannot stop abruptly where our microscopes cease to be of use. Here the vision of the mind authoritatively supplements the vision of the eye. By an intellectual necessity I cross the boundary of the experimental evidence and discern in Matter .... the promise and potency of all terrestrial life.'

[The existence of such low types of organic life as, for instance, Amoebae, the mere speck of jelly which still digests, lives, and reproduces itself, shows in what low forms animal life, if it really be such, can exist. In an interesting paper on 'The Border Territory between the Animal and the Vegetable Kingdoms,' Professor Huxley has tried to show how near in some cases vegetable comes to animal life. Microscopical research more and more confirms the generalization of Schwann, that 'a fundamental unity of structure obtains in animals and plants;' and that the structures of the bodies of both, however diverse in fabric, 'result from the metamorphoses of morphological units (termed *cells* in a more general sense than that in which the word 'cells' was at first employed), which are not only similar in animals and in plants respectively, but present a close fundamental resemblance when those of animals and those of plants are compared together.' Again, Huxley 'knows of no test by which the reaction of the leaves of the Sundew and of other plants to stimuli, can be distinguished from those acts of contraction following upon stimuli which are called "reflex" in animals. .... It must be allowed to be possible that farther research may reveal the existence of something comparable to a nervous system in plants.' Huxley illustrates his argument by quoting the history of the fungus (*Peronospora infestans*) which is the cause of the potato disease. This fungus develops spores which sometimes germinate in the ordinary way, but more commonly break up into six or eight zoospores or living organisms, which at first swim about in the moisture on the surface of the leaf, and then become quiet and germinate like a vegetable in the tissues of the plant. Thus, whether we start from the animal-side or from the plant-side, 'such an insensible series of gradations leads to the monad, that it is impossible to say at any stage of the progress—here the line between the animal and the plant must be drawn.' Thus Huxley concludes, in certain cases the problem whether an organism is an animal or a plant may be essentially insoluble.]

With considerable force Tyndall next argues that the phenomena of crystallization show that matter possesses a structural power. The polarity of magnetism appeals to the senses, and gives a basis for the 'conception that atoms and molecules are endowed with definite attractive and repellent poles, by the play of which definite forms of crystalline architecture are produced. *Thus molecular force becomes structural.* It required no great boldness of thought to extend its play into organic nature, and to recognize in molecular force the agency by which both plants and animals are built up.' He had formerly used the formation of ice as a simple illustration of this process. When solid crystals of ice are produced—

'By their own constructive power, molecule builds itself on to molecule with a precision far greater than that attainable by the hands of man. . . . . Imagine the bricks and stones of this town of Dundee endowed with locomotive power. Imagine them attracting and repelling each other, and arranging themselves in consequence of these attractions and repulsions to form streets and houses and Kinnaird Halls,—would not that be wonderful? Hardly less wonderful is the play of force by which the molecules of water build themselves into the sheets of crystals which every winter roof your ponds and lakes. . . . Latent in every drop of water lies this marvellous structural power, which only requires the withdrawal of opposing forces to bring it into action.'[1]

In a lecture delivered more lately,[2] Tyndall has expanded the

[1] 'Fragments of Science.' Matter and Force, a Lecture to the Working Men of Dundee. Pages 82 and 85.

[2] On the subject of 'Crystalline and Molecular Forces.'

same thought. After showing some experiments to illustrate the forces of crystallization he said, in concluding,—

'Everywhere, throughout our planet, we notice this tendency of the ultimate particles of matter to run into symmetric forms. *The very molecules appear inspired with the desire for union and growth;* and the question of questions at the present day is—and it is one which I fear will not be solved in our day, but will continue to agitate and occupy thinking minds after we have departed—how far does this wondrous display of molecular force extend? Does it give us movement of the sap of trees? I would reply with confidence, 'Assuredly it does.' Does it give us the beating of our own breasts, the warmth of our own bodies, the circulation of our own blood, and all that thereon depends? This is a point on which I offer no opinion to-night.'

This is a partial outline of what Martineau[1] calls 'the *new* book of Genesis.'

To this address Dr. Martineau replied in a pamphlet which was indeed unsatisfactory as a criticism, for it did not in any respect profess to be a complete one. It was but an expression of strong disagreement with the other's conclusions. In a later article Tyndall has replied to this, and endeavours to show that we need not go outside of Matter to find the power which produces life and consciousness. The same structural force which produces the beautiful ice-crystals and, as he believes, makes the sap circulate in the trees, can in a different mode of action make the blood circulate in a human body, and call into being the consciousness which we feel ourselves to possess. In this second paper Tyndall develops this thought far more fully and uses that graphic power of illustration which he can command so readily. His treatment of the subject shows an admirable power of exposition; indeed it could hardly be more clear and vivid than it is.

Imagine Matter in the shape of vapour rising off the

[1] 'Religion as affected by Modern Materialism.' By James Martineau. 1874, p. 8.

heated surface of the Caribbean Sea; trace it from the tropics flying northwards and eastwards across the Atlantic till it reaches Europe and the Alps. Here a wonderful transformation takes place. The cold air condenses the vapour to water in the form of snow-crystals, perfectly accurate in geometrical form and exquisitely beautiful: and these crystals are Matter in another form. What has occurred here? he asks. Has an 'imponderable formative soul'—by this Tyndall implies any possible outside creative Power—entered into them from without and built each particle into its proper place? or, on the other hand, has the formative power worked from within, and was it locked up within the drops of water from the very first? Emphatically it was the latter. Again, taking an example from the world of life, it is Matter, mere Matter, with no addition, which grows up from the acorn into the stately oak. What a wonder, then, results when, from the interaction of the acorn, the earth and the sunlight, each of which is nothing but Matter, arises the massive oak, shaking its countless twigs in the breeze. And here is verily what we call life, sprung from the interchange of mere Matter. And Tyndall, apparently somewhat presumptuously, says, 'It will be seen that I am not in the least insensible to the wonder of the tree; nay, I should not be surprised if, in the presence of this wonder, I feel more perplexed and overwhelmed than Mr. Martineau himself.' This statement seems to need some justification, and accordingly he uses most admirable illustrations to show how keenly the scientific man realizes the wonderfulness of vegetable growth. He refers first to an experiment in which the 'music of a piano is transferred from its sound-board through a thin wooden rod, across several silent rooms in succession, and poured out at a distance from the instrument.' All the vibrations and sub-vibrations of the strings—more than a hundred occurring at the same time—are crowded together and poured through this narrow bit of wood. There is no confusion, no note is lost, but all are shaken forth into the air in a distant chamber through a second sound-board against which the end of the rod presses. 'Thought,' he says, 'ends in amazement when it seeks to realize the motions of that rod as the music flows through it.' Look next at one of those exotic ferns, the branches of which are sometimes hardly thicker than a pin, and quite leafless for a foot and more. But at the end of this bare thin stem bursts out a mass of fronds almost large enough to fill the arms. All the shocks and pulses and other vital actions which produce that luxuriant foliage must flow through the insignificant stems. 'We stand here upon a higher level of the wonderful. We are conscious of a music subtler than that of the piano passing unheard through these tiny boughs,' and issuing in the splendid cluster of leaves. And realizing this so keenly, he claims to feel the wonder and beauty of this sight as strongly as his opponent, and indeed more so.[1]

Again, in answer to Martineau's somewhat haughty criticism, that, according to Tyndall's conception, 'Matter starts as a beggar without a rag and turns up as a prince when great undertakings are wanted,' he replies, 'Not so. Mysterious and incomprehensible as it is, this wealth of power is within it from the first. It is Matter and only Matter, with no addition, which grows up from the seed into the oak, from the embryo into the perfect animal or man.' And, after referring to the organization of the human being and the marvels implied in the formation of the human eye and ear, before the infant's birth, he concludes, 'Matter I define as that mysterious thing by which all this is accomplished.'[2]

This whole article, as well as Tyndall's former one, illustrates,

[1] Tyndall has told us a part of the Story of the Flower or Tree which is wonderful indeed, yet it is but a part. Wordsworth can tell us far more.

[2] We may compare Spinoza, who says, with reference to body and mind, 'Yet no man has hitherto determined what body [or Matter] can do, Matter that is, solely by the laws of its own nature, in so far as it is mate-

in particular, the unwillingness of many scientific men to admit a Power acting on Matter 'from without,' a God who 'orders the universe from a position outside it all.' Goethe before used the same expression in that poem where he says,

> Was wär' ein Gott der nur von aussen stiesse? . . .
> Ihm ziemt's die Welt im Innern zu bewegen.

'What were a God who only impelled the world from without!' But then the question arises—What does *without* mean, and what does *within* mean? That would be indeed hard to define. But Goethe's line has a distinct enough force, and that is, 'How mean is that conception of God which makes him act only at intervals, by fits and starts, not constantly both here and everywhere!' The working from within, which is according to Goethe the nobler conception, appears to imply an ever-working, everywhere-present God. Tyndall's omnipresent Formative Power, residing in Matter—however he may have narrowed the notion—does point, if but dimly, towards the omnipresence of the Divine Energy, as religious thought loves to realize it. It need not be forgotten that the notion of a Divine Mechanist, who sometimes from far away interferes with the laws of Nature, while these at other times work in entire self-sufficience, is fully as repugnant to the greatest religious thinkers—Augustine, for example—as it can be to any scientific man.

Since this reply of Professor Tyndall's appeared, the battle has been fought fully out; this great controversy of our day has been carried out on both sides to its utmost issues, and a crowd of spectators have watched its progress with an interest of more than curiosity. Some who felt old truths almost slipping from beneath their feet, and, along with this, a unique sensation of universal unsteadiness and falling, like that of men in an earthquake, when the solid earth which they have

rial, and what it cannot do without direction from mind' ('Ethica,' Pars iii., Prop. 2).

known all their lives, and which has ever been firm under their step, even the earth begins to be unsteady and shake under them—these have felt warm gratitude to the practised champion of philosophy who calmly accepted the challenge so confidently flung down, and whose strength and courage won a perfect victory. If some at first feared lest insensate Matter should soon become lord of all, they found when they took courage and closely grappled with the terror, that it fell limb from limb in their grasp.

Professor Tyndall—for we refer to the controversy between him and Martineau, a controversy which will be admitted to be on both sides a fairly representative one—might indeed have known that the antagonist with whom he wrestled put forth but half his power. Martineau's first pamphlet was but an expression of strong dissent. More lately, however, he has examined Tyndall's arguments most calmly and fully, and has discussed them one by one, meeting them in every case, so far as we can judge, with a singular fairness. The veteran in the domain of mind, when roused into fuller action, has met his antagonist on every point of attack, and has cast him so heavily that he cannot rise. Not by any means that we rejoice to see the representative of science defeated. While we think his conclusions entirely unphilosophical, we admit that Tyndall has stated much that is unquestionable, much that makes the world more wonderful to us than before. We do not wish to dwell on his somewhat 'petulant' statements—that Martineau 'rashly kicks away the only philosophic foundation on which it is possible for him to build his religion,' or that the character of his creed is 'entirely subjective,' or the too-confident heroics on the 'comfort of belief.' It would be easy to criticise such words, but we have been much more anxious to follow Tyndall to his own ground and, coming at once to the central question on which the whole controversy turns, to state what powers he attributes to Matter, and to show what reasons he can give, just from his own point of view, for assigning these. Here we have found that he has much to teach us, much that but for him we might have missed entirely.

Martineau's criticism on all the points before quoted is admirable. He remarks in the first place that Tyndall, while treating of and explaining the methods of Nature, entirely drops out of thought the Power which works along these methods. This criticism seems to us thoroughly just. He says of the series of illustrations which lead up to Tyndall's aphorism—

'"Matter I define as that mysterious thing by which all this has been accomplished," *i.e.*, the whole series of phenomena from the evaporation of water to self-conscious life of man. Need I say that such a proposition is no definition and dispenses with all proof, being simply an oracle, tautologically declaring the very position in dispute, that matter carries in it "the promise and potency of terrestrial life." The whole of the picturesque group of descriptive illustrations which lead up to this innocent dictum are only an expansion of the same *petitio principii;* they simply say, over and over again, the force immanent in matter *is* matter—they are identical. . . . This is not a process of reasoning, but an act of will. Nothing can be less relevant than to show (and nothing else is attempted) that the forces of heat, of attraction, of life, of consciousness, are attached to material media and organisms, which they move and weave and animate: this is questioned by no one.'

And again, 'The very story of saline crystals and ice-stars and fern-fronds and human birth, which Professor Tyndall tells in order to exclude it [*i.e.*, causal Will], is to me a continuous report of its agency and laws. He asks, "What else is there here than matter?" I answer, the *movements* of matter with their disposing and "formative power," the attracting and repelling energies which, *dealing with* molecules and cells, *are not* molecules and cells.'

Thus we come to see that science deals only with sequences and succession of phenomena—not with dynamic ideas—not with causality. So, imagining the case of a perfect observer, able to follow all the changes of external bodies, 'though the whole objective world has been laid bare before him, and he has read and registered its order through and through, he has not yet, it will be observed, alighted on a single *dynamic* idea. . . . The whole vocabulary of causality may absent itself from such an observer. . . . He might, as Comte and Mill and Bain truly contend, command the whole body of science, without ever asking for the origin (other than the phenomenal prede-decessor) of any change.' So the scientific text-books wrongly speak of 'Energies'—they can only explain the succession of movements, 'accomplished, prevented, modified.' Science cannot tell us *what* it is which does the work—*what* it is which produces these movements of matter.

More recently Dr. Carpenter has discussed this very subject of the relation of Matter to Force in an article entitled, 'The Force behind Nature.'[1] Dr. Carpenter attempts to show that, through what he proposes to call our Force-sense, we have a 'direct cognition of Force;' that our 'cognition of Force is quite as immediate and direct as our cognition of motion,' and that it is one of those 'universal ideas which belong to every human mind.' One illustration of his is apt enough for this part of our inquiry, and in other ways is extremely suggestive. Man's position as a student of Nature is compared by him to that of a man in a cotton-factory, whom he supposes to enter it in order to study the working of the machines, *without knowing in the least what is their moving power.* His first result of observation is to classify the different kinds of machines, each according to the work it does. Next, concentrating his attention on any one, he might not unnaturally suppose the machine to be self-moving, and 'he might thus attribute to each kind of machine an *inherent power* of carding, roving, drawing, spinning, or weaving, as the case may be.' He would next observe that from time to time one

[1] 'The Modern Review,' January, 1880.

or other machine stops, and then goes on again; this stoppage follows when the man working at it disconnects its axis from a leather band, and this leather band, he notices, runs over a great longitudinal shaft at the end of the building. This connection with the shaft he sees to be essential to the working of the machine, but as yet he does not in the least understand *why* it is essential. The next fact which he observes, however, conveys an entirely new idea to his mind. If we suppose him to lay hold of the band, which goes between the shaft and the axis of any one machine, he at once becomes conscious, through his force-sense, that the band is an instrument conveying power, and that, so far from any one machine having an inherent power of movement, the *source* of its power lies in the shaft. Finding out that every machine in the factory derives its power from the same source, he next inquires, Has the shaft itself an inherent power of motion? or does it derive that power from any ulterior source? While thus studying the working of the machines, suddenly he sees all the machines stop, and the shaft cease to revolve; and after the interval of an hour both shaft and machines resume their motion, and yet he 'is certain that no agency visible to him' has had any concern either in their stoppage or in their renewal of motion. Thus he might feel justified in concluding that the shaft possesses an inherent potency of motion, and that there is no more to be learned about it. But, not wishing to leave anything uninvestigated, he goes round *to the other side of the wall.* There he finds that one end of the shaft comes through it, and is set in motion by a steam-engine.

Of course, the force of the engine is derived from coal, and, if we go far enough back, from solar radiation. If, then, we inquire into the source of the sun's energy, in any case 'we come at length to a wall, to the other side of which we seem at present to have no access.'

But, continues Dr. Carpenter, *Is* there no other side? Does not the whole preceding inquiry show how unsatisfactory is any inherent 'potency' of Matter as the ultimate explanation of the existing Kosmos? 'If we think the man foolish who supposes the main shaft of a cotton-mill to turn *of itself*, merely because he sees it apparently end in a wall, which conceals from him the source of its motive power, are we not really chargeable with the like folly if we attribute self-motion to the ultimate molecules of Matter, merely because the Power that moves them is hid from our sight?'

The misconception of confusing 'Law' and 'Force,' which Dr. Carpenter exposes, has become so widespread that we may be excused for quoting his little parable thus fully. Martineau next points out,—Observation shows us that all Forces are One. If you watch any one force, it may be seen passing into another. 'Now it is mechanical energy, in a minute it will be heat; if a tourmaline is near, it will turn up as electricity; and so on, for no part of the cycle is closed against it. You look, in short, upon a row of masks, behind which the "unknown power," slipping from one to another with magic agility, seems to multiply itself, but is found on closer scrutiny never to quit its unity.'

So the plurality of forces disappears. The forces are seen to be mere disguises of some unity, and we next inquire, How is that One Power related to its many masks? Is it indifferently related to all, but like none of them? or are the phases not on equal footing, but consecutive? and if so, is the lowest or the highest to be held the oldest? Three answers may be given. At present we can only refer to Martineau's most close discussion of each. It is needless to say what important consequences are implied in the answer that may be chosen.

Next, we cannot help inquiring, What is this Power? But Science cannot tell us, for it cannot discern Power! To find out its nature, we must look within ourselves. What makes us at the sight of phenomena ask for a power? It is the active side of our own nature. We are conscious that we ourselves exercise power. This is the one thing that we immediately know. We call this exercise of power *living Will.* Our whole idea of *Power* is identical with that of *Will*, or deduced from it. So by a mental postulate we recognize causality in Nature. The law of causality in its intuitive form is this—phenomena are the expression of living energy. Till this intuition is disproved, the One Power stands as the Universal Will. (If you take away this from dynamic conceptions, you are again reduced to co-existences and successions.) Nor is this notion so repugnant to all scientific men as it is to some at present. Herschel says that 'it is our own immediate consciousness of *effort* when we exert force to put matter into motion, or to oppose and neutralize force, which gives us this internal conviction of *power* and *causation*, so far as it refers to the material world, and compels us to believe that whenever we see material objects put in motion . . . or deflected if already in motion, it is in consequence of such an *effort somehow* exerted, though not accompanied with our consciousness,'[1] and others speak to the same effect.

In a later article,[2] not referred to by Martineau, Herschel has worked out the same thought more fully—

'In that peculiar *mental sensation*, clear to the apprehension of every one who has ever performed a voluntary act, which is present at the instant when the determination to do a thing is carried out into the act of doing it (a sensation which, in default of a term more specifically appropriated to it, we may call that of *effort*) we have a consciousness of immediate and personal causation which cannot be disputed or ignored. And when we see the same kind of act performed by another, we never hesitate in assuming for him that consciousness which we recognize in ourselves.'

[1] 'Treatise on Astronomy,' chap. vii., § 370. 1833.

[2] 'On the Origin of Force' in 'Popular Lectures on Scientific Subjects,' p. 461. By Sir John Herschel. 1873.

The next step in the way of generalization, Herschel continues, is 'a flight rather than a step,' yet it is 'one that forces itself on our thoughts with ever-increasing cogency.' Whatever event takes place in the material world, we either find to be ultimately resolvable into some change occurring in material substance, or we endeavour to trace it to this. 'In every such change we recognize the action of *Force.* And in the only case in which we are admitted into any personal knowledge of the origin of force, we find it connected (possibly by intermediate links untraceable by our faculties, but yet indisputably *connected*) with volition, and by inevitable consequence with *motive*, with *intellect*, and with all those attributes of mind in which personality consists.'

We have elsewhere quoted (p. 116) from the same article another most suggestive passage as to the method in which we may suppose Will to act on Matter.[1]

Herschel's paper greatly supplements Martineau's treatment of this point, which, we cannot help thinking, is somewhat hasty.[2] Yet the thought is clear—namely, it is a mental postulate to refer Power to Will. Any causal power other than Will, is, he says, 'absolutely out of the sphere of thought.' To this point we shall return. But here we cannot help quoting one other passage, referring to the relation between God and the world.

Martineau suggests that inorganic matter and lower animal life represent as it were the habits of the universe, habits carried on somewhat in the way that men do certain acts—mechanically—while the Indwelling Mind concentrates its aim on the natures that resemble itself. It is in planning new

[1] In the same article (pp. 468-73) Herschel discusses the question, How does the scientific principle known as the 'Conservation of Energy' bear on this question of Will and its power to originate Force? Does that principle 'stand opposed to any, even the smallest amount of arbitrary change in the total of "force" existing in the universe?'

[2] He states the same position more fully in his 'Essays,' vol. ii., pp. 185-9.

methods of work, he says, that the human mind exercises its highest functions. When it has established a definite method of working, the strain is relaxed: a habit is formed that can almost execute itself, so that the mind can press on to new schemes, while still at the same time the same mind carries on the old usages—

'Does anything forbid us to conceive similarly of the kosmical development; that it started from the freedom of indefinite possibilities and the ubiquity of universal consciousness; that as intellectual exclusions narrowed the field, and traced the definite lines of admitted movement, the tension of purpose, less needed on these, left them as the habits of the universe, and operated rather for higher and ever higher ends not yet provided for; that the more mechanical, therefore, a natural law may be, the further is it from its source; and that the inorganic and unconscious portion of the world, instead of being the potentiality of the organic and conscious, is rather its residual precipitate, formed as the Indwelling Mind of all concentrates an intenser aim on the upper margin of the ordered whole, and especially on the inner life of natures that can resemble Him.'

The track we are following has brought us here to a point from which a noble prospect bursts into view. These words open up a whole new range of thought that surprises us like a glimpse 'from magic casement' into some untrodden land.

Tyndall's theory of 'organic matter,' which with its supposed results Martineau has throughout been combating, points plainly to but one ultimate result, nor does Tyndall shrink from stating this. 'From this point of view,' he says, 'all three worlds (the inorganic, the vegetable, and the animal) constitute a unity in which I picture life as immanent everywhere;' and this life, he says, may possibly be but 'a subordinate part and function of a *higher life*.' This is indeed a sublime conception, says Martineau. We have no scale of life, higher than human, in the world. Scale of height above this is only in degrees of the intellectual and the moral. So if that 'higher life' exist at all, we must think of it as transcendent Mind and Will. Yet to his so-called 'higher life' Tyndall dares not give the predicate 'Mind,' or apply the pronoun of personality. On what scale then, asks Martineau, is it 'higher'? If not on the intellectual and moral, then there is in man what rises above it, 'for the power of attaining truth and goodness is ideally supreme. If Tyndall can reveal to us something higher than Mind and free Causality, let us by all means accept and assign it to God. But to profess this, one would have, I think, to be something more than human. Else how could one grasp its conception? how look higher up than the level of Reason. If that "higher life" speak to us in idea at all, it can only be as Perfect Reason and Righteous Will. *Those who find this type of conception not good enough, do they succeed in struggling upwards to a better?* . . . . Rather, I should fear, they *droop and sink into the alternative faith of blind force.*'

Scientific thought seems at present to move strongly in the direction which Tyndall has stated, namely, Life is immanent in Matter, and no Directive Power can act from without on the world of Matter. We have shown already that bare and naked atheism is not the only possible result of such a theory. One solution of the difficulty, and a very natural one, is offered in the essay entitled, 'The Mystery of Matter,' by Mr. Picton,[1] a writer who has met scientific discovery with the frankest welcome, while he holds its philosophical and religious bearings with a very different grasp from Professor Tyndall's. Mr. Picton endeavours to show that the ordinary conception of atoms, as indivisible particles which occupy space exclusively, is untenable. If this opinion be accepted, how can 'two substances—like oxygen and hydrogen—produce a third so utterly unlike both as water?' Why not rather think of

[1] 'The Mystery of Matter,' by J. A. Picton. Second edition. 1878.

the atoms as others do of the interspaces between them, and regard them as the mere 'phenomena of force.' Picton sees 'strong reasons for believing' that matter is but accumulated 'centres of force.' 'We may suppose these centres capable of interpenetrating one another, and of thus producing an entirely new mode of force, or, in common language, a new substance.' The Atomic theory, pure and simple, 'first denies and then is compelled to assert the dissociation of matter and force.' The ordinary conception of dead atoms, of an 'unliving substance' called matter, Picton utterly rejects. 'The notion of a dead substance, foreign to and incommensurable with spiritual being,' is a mere 'spectre' which is 'entirely the creation of false inference.' We are certain of only one thing, namely, the existence of life, our own or another mode of life. 'We know that life is, but we do not know that anything else is.' Our notion of matter as a dead substance apparently comes from our experience of the 'physical laws which bound the efforts of our will.' We naturally conceive of force, whether put forth by ourselves or in nature, as effort. We observe that these laws act in sequences, and since these sequences occur with 'certainty and unfeeling regularity,' this appears to us ground for excluding from them 'the immediate action of will or conscious purpose.' We abstract these and there remains the 'really irrational conception of unliving effort.' But mere force will never prove a solution of the mystery of matter. 'Both forces and forms, far from lending themselves to "gross materialism," rather fascinate us with their shadowed hints of a mystery behind them both, far mightier than our will, *and I will dare to add, more keenly living than our life.*' This is why landscape has the power to touch us so deeply.

Mr. Picton is willing to concede to Professor Huxley that 'force is as much a function of matter as motion is.' But at the same time, as he remarks, 'the whole significance of the concession depends upon the meaning that we attach to the word "matter."' And then he propounds with elaborate illustration the theory that matter 'is in its ultimate essence spiritual.' Matter is the 'elementary phenomenal definition,' such as our consciousness apprehends, 'of a universal spiritual Power.'

Thus he claims to have 'gone right through materialism, and come out at the other side, where it merges into pure spiritualism.'

However mistaken his final conclusions may be, Mr. Picton's theory of Matter is at least a grand poetic dream. Not a few passages of his book intoxicate the reader. We cannot here analyze the argument by which he defends his view of Matter, but we may remark that the question of Mind and Matter cannot be quite so simply settled as Mr. Picton appears to believe. His dilemma is not close enough. The world might be conceived, instead of being the 'elementary and phenomenal,' to be the *partial* and imperfect 'manifestation' of the One Power—a side-stream returning into, but ever flowing apart from, its source, and which it is possible for us to know only while it is separated from that source, somewhat like men dwelling where they can see only a side channel, but never the great river from which its waters come. If this were possible, Matter and Mind need not be viewed as two utterly discordant substances. Poetically beautiful, morally earnest, and in many respects most truly religious as Picton's form of Pantheism is, and it can be such only because there is much of truth mixed with it, yet in reference to this subject—in trying to realize the relation of the world to God—we shall find something far more helpful, far closer to fact, in Martineau's suggestion that inorganic matter and lower animal life may represent, as it were, 'the habits of the universe' carried on 'mechanically.' Thus we complain that Nature is unsympathetic and heedless and we feel that an infinitely deeper Divine communion is possible through the voice speaking within than through the world outside, because 'the Indwelling Mind of all concentrates an intenser aim on the upper margin of the ordered whole, and especially on the inner life of natures that can resemble Him.'

We have now surveyed the controversy of Matter over all the disputed ground, and what is the result? Is it that the whole world has become disenchanted and dead? or do not the consequences far rather tend the other way, and has not the earth our dwelling-place, with all its mysteries of life and growth, become more wonderful than we before thought it? Is not the working of the One Power on Matter something more strange and beautiful than we were wont to think? Matter in every shape—from the dead mass, throbbing with countless unseen movements which mind can barely imagine, to the little company of snowdrops hanging their heads where a month before was bare black ground—is yet more mysterious than before.

CHAPTER VI.

EPICUREAN PSYCHOLOGY.

ONE sometimes wonders how any man who lives and feels his own life keenly can ever believe that a number of atoms, even granting countless ages and the usual formula, could of themselves produce human beings. An acute living thinker says emphatically that nobody ever actually and really believed such a thing. If he once tried to realize it, all he means by such a profession is that no one can get any further than this in the way of explaining the problem. Probably this is true. At the same time Lucretius did come very near to so believing; he schooled his mind so deliberately into this habitual attitude that he felt quite convinced logically that atoms in motion were amply potent to produce the world and man. At the same time, we think it possible that some men now living are really still nearer this than Lucretius, only they are not so outspoken, not so sternly uncompromising in expressing their convictions as he was. One cannot help being curious to know by what reasoning Lucretius actually reconciled himself to such a belief, and how he accounted for the purely atomic origin of Consciousness. In considering his attitude we may find it more than once inconsistent; but in examining it at all, the only way worth while is to let him paint himself as he is. Singularly enough, and if so in the sheerest self-contradiction, we may find even him compelled at last to admit the existence of something more than Matter.

After having thoroughly proved to his own satisfaction the existence of the atoms, their indestructibility and eternal motion, Lucretius next endeavours with much emphasis and ardour to demonstrate that these atoms are utterly dead.[1] They are not living like weak flesh and blood, for living things are necessarily perishable, but not so the atoms. Moreover, life implies internal motions of pleasure and pain, but the atoms are perfectly hard and solid, as has been proved; there is no room in them for such movements. Besides, if the atoms possess life, how can many living atoms form a single body with one vital sense?

How then does Lucretius account for these little particles producing living beings? He commences to answer this by explaining that it is a mere prejudice to suppose that life can come only from life. We think that Matter cannot produce life, and that is natural, because we think of Matter as we see it every day in the shape of masses of stone and logs of wood, and these substances indeed could never produce life, however much they were to be jumbled together. But the fine minute atoms can mix together in a very different way. Matter in this shape is quite competent to do all that he claims. Then, of course, he repeats his formula of evolution,—'It matters much with what others and in what position the same atoms are held together,' and so on. To produce living things as distinguished from dead objects, the atoms only require to be specially minute, of special shapes, and to have fallen into special arrangements and mutual motions.[2]

[1] ii. 865-1022.

[2] nimirum lapides et ligna et terra quod una
mixta tamen nequeunt vitalem reddere sensum.
illud in his igitur rebus meminisse decebit,
non ex omnibus omnino, quaecumque creant res,
sensile et extemplo me gigni dicere sensus,
sed magni referre ea primum quantula constent,
sensile quae faciunt, et qua sint praedita forma,
motibus, ordinibus, posituris denique quae sint.
quarum nil rerum in lignis glaebisque videmus. ii. 889-97.

Pleasure is but an orderly motion of the atoms, pain is felt when they move in disorder. Life is but a 'mode of motion' on the part of the atoms. This stipulated order in their arrangements and motions is indeed important. Lucretius insists upon it: you cannot have life without it, but it implies no arranging Intellect. The existence of life as distinct from death means merely that a multitude of little particles have changed their position. So we see the earth after rains produce numbers of creeping things, just because its particles have fallen out of their old combinations, resulting in dead matter, into the new ones necessary to produce life.[1] In the same way a strong enough cause can in a moment stop these motions of the atoms, and then death takes place. The difficulty of the origin of sensation he hardly realizes. Body and soul cannot, he says, feel apart from each other, but their mutual motions 'kindle' sensation which 'bursts up like a flame'[2] between them. Again, he asks, feeling plainly that he has got a very strong instance in confirmation of his theory,—What of a man in a swoon? if activity of the senses and consciousness is a proof that a principle of life exists apart from the body, how is it, when a man receives a heavy blow, that for a time he loses consciousness and then returns to life? Where was the true self during that interval? It made no sign. Did it die and come to life again when the man recovered? And he gives what he feels to be a very plausible explanation. Life is but a certain orderly motion of the atoms;

1 et tamen haec, cum sunt quasi putrefacta per imbres,
vermiculos pariunt, quia corpora materiai,
antiquis ex ordinibus permota, nova re
conciliantur ita ut debent animalia gigni. ii. 899-901.

2 nec congressa [sc. materies] modo vitalis convenienti
contulit inter se motus, quibus omnituentes
accensi sensus animante in quaque cientur. ii. 941-3.

This comparison of sensation to a fire-flash between flesh and soul is found also at iii. 335-6, and iv. 927-8.

when this motion is greatly disordered by a blow, life almost passes away; but the remnant of the proper vital motion prevails over the disorder and recalls the atoms to their proper motion, and so the return to consciousness takes place. If life departs because this motion is disturbed, and returns when it is restored, then life is nothing more than a certain motion of the atoms. In this way, as he says, Nature produces living bodies from Matter in the shape of food, much in the same way as she turns wood into fire.[1] 'Nature' (by this meaning the habits of the atoms) is little more here than a figure of speech, though, as we have seen, it is easy to find a modern parallel to the statement that living bodies are only the product of 'Matter and Nature,' that is, of 'Matter and the habits of Matter.' At present we may only remark that Lucretius has strong reasons for labouring so eagerly as he does to prove Matter to be 'utterly dead.'[2] He must have seen one weak point in his system, or probably he cannot be said to have *seen* but only to have *felt* its presence, and though his present argument is entirely powerless to answer the objection, it is probably aimed towards it. But this will become more clear afterwards. Plainly, Lucretius's theory of Matter is such that henceforth no more divine agents were needed to create living things, or to act in any way on Nature *from without.* 'Nature,' he says, 'is seen to do everything alone and entirely of her own accord'—

> Quae bene cognita si teneas, *natura videtur*,
> libera continuo, dominis privata superbis,
> *ipsa sua per se sponte omnia dis agere expers.*

1 ergo omnes natura cibos in corpora viva
vertit et hinc sensus animantum procreat omnes,
non alia longe ratione adque arida ligna
explicat in flammas et in ignis omnia versat. ii. 879-82.

2 seminibus carentibus undique sensu. ii. 990.
haut igitur debent esse ullo praedita sensu (sc. primordia). ii. 972.

Pursuing his aim, the deliverance of mankind from superstition and fear, Lucretius proceeds in his Third Book to explain the nature of the soul, and to show that it perishes along with the body. Thus the fear of torments after death is to be done away with. The soul is material, for without touch how can it direct and act on the body? It is closely united to the body; just as it is hard to separate the odour from a lump of frankincense without destroying it, so it is impossible to part soul and body without destroying both. The soul does not live 'in a den of its own,'[1] but is spread all over the body. It 'grows along with the body, together with its members, within the very blood.'[2] The atoms forming it are vastly different from those composing flesh and bone. They are 'exceedingly small, smooth and round';[3] how much Epicurus thinks to be implied in such special fineness of atomic composition, we have seen already.[4] The soul-atoms are also fewer and at far greater intervals than those of the body.[5] The bulk of the soul is exceedingly small compared with that of the body. As Gassendi[6] expresses it, following Lucretius, 'If you can conceive

[1] velut in cavea per se vivere solam. iii. 684.
Lucretius is here refuting the notion that the soul enters into the body after the latter is fully formed, at the instant of birth, instead of growing along with it from the first. In the former case, he says, the soul must of necessity dwell by itself in a distinct part of the body.

[2] uti cum corpore et una
cum membris videatur in ipso sanguine cresse. iii. 682-3.
Compare iii. 445-58.

[3] iii. 203-5. According to Philoponus, De An. c. 6, Democritus considered the round atoms to be the smallest. We may regard it as certain that Epicurus did the same.

[4] See p. 59.

[5] nam cum multo sunt animae elementa minora
quam quibus e corpus nobis et viscera constant,
tum numero quoque concedunt et rara per artus
dissita sunt dumtaxat. iii. 374-7. Compare ll. 276-8.

[6] 'Adeo proinde, ut punctum sit prope loculusve exilissimus in quem recipi possit anima, si totum fingas conglomerari.'—'Epicuri Philosophia,' Leyden, 1675, vol. i., p. 404. Lucretius himself says,—

the whole of the soul to be gathered into one mass it would occupy a mere point almost, or the very tiniest space.'[1]

The *anima*, the soul and life (in which Lucretius often includes the mind, *animus*), is composed of four different substances,—wind, heat, calm air, and a fourth substance.[2] The soul is *anima*, formed of the first three only, while the mind (*animus*) is composed of these and of the fourth substance as well.[3] This is 'entirely without name'—quite beyond definition. It is the finest and most nimble of existing things, and nothing else is formed of such small and smooth atoms, for such only can accompany the marvellous swiftness of the mind in thinking and imagining.[4] This is the innermost and

haec quoque res etiam naturam dedicat eius,
quam tenui constet textura quamque loco se
contineat parvo, si possit conglomerari. iii. 208-10.

the proof of which is that the body appears in no whit either smaller or lighter after death than it was before.

[1] The exceeding fineness in texture of the mind as compared with the body is implied in Lucretius's doctrine of mental images. These, though beyond expression thin and fine, are perceived directly by the mind. Moreover, a single such image is sufficient to rouse mental consciousness, though of the far grosser images which enter the eye, a constant succession is needed to produce sight. Any one such mental image 'readily moves the mind by a single stroke, for the mind itself is fine and wondrously nimble.'

facile uno commovet ictu
quaelibet una animum nobis subtilis imago;
tenuis enim mens est et mire mobilis ipsa. iv. 746-8.

[2] iii. 231-57.

[3] So Munro has pointed out. Lange ('Geschichte des Materialismus,' vol. i.) seems not to identify the fourth substance with the *animus*.

east omnino nominis expers;
qua neque mobilius quicquam neque tenuius exstat,
nec magis e parvis et levibus est elementis. iii. 242-4.

Here, as so often, Epicurus treads in the footsteps of Democritus, as we see from a passage in Aristotle's 'De Anima,' i. 405: ψύχην μὲν γὰρ εἶναι ταὐτὸ καὶ νοῦν, τοῦτο δ' εἶναι τῶν πρώτων καὶ ἀδιαιρέτων σωμάτων (the atoms), κινητικὸν δὲ διὰ μικρομέρειαν καὶ τὸ σχῆμα· τῶν δὲ σχημάτων εὐκινητότατον τὸ σφαιροειδὲς λέγει· τοιοῦτον (that is to say, 'most mobile') δ' εἶναι most vital portion of the soul. It is the guiding principle at whose supreme bidding the soul moves; it is 'the very soul of the whole soul'—

> Atque anima est animae proporro totius ipsa.

It alone is the origin of will and thought. Sensation and motion begin from it.[1] 'It first transmits sense-giving motions over the frame; for it is the first to be stirred, from the smallness of the atoms which compose it; next the heat and the unseen element of wind receive the motion, then the air, and then everything is set in movement,' and all over the body, in flesh, blood, and bone, we feel either pleasure or pain.[2] And in a striking passage (ii. 261-93) Lucretius shows how motion begins from the heart, the seat of will, and is thence transmitted

τὸν τε νοῦν καὶ τὸ πῦρ. We may observe that as to the composition of the soul, Epicurus diverges very widely from his master. Democritus considered the soul to be formed *entirely* of atoms the same in shape as those of fire—that is to say round, but arranged in a different order, and, probably, smaller also than the fire-atoms. Democritus calls the soul not, as Zeller renders it, fire, but *πῦρ τι*, 'a kind of fire.' ('De Anima,' i. 403.)

[1] This point is important. On it the evidence is clear. Plutarch ('De Plac.,' iv. 3), says—*Επίκουρος κρᾶμα* [sc. *εἶναι τὴν ψύχην*] *ἐκ τεσσάρων, ἐκ ποιοῦ πυρώδους, ἐκ ποιοῦ ἀερώδους, ἐκ ποιοῦ πνευματικοῦ, ἐκ τετάρτου τινὸς ἀκατονομάστου, δ ἦν αὐτῷ αἰσθητικὸν* (quae ei erat vis sentiendi, as Dübner translates). The same is found in Stobaeus (i., p. 798), who continues, *τὸ δ' ἀκατονόμαστον τὴν ἐν ἡμῖν ἐμποιεῖν ἀίσθησιν· ἐν οὐδενὶ γὰρ τῶν ὀνομαζομένων στοιχείων εἶναι ἀίσθησιν.* So also Lucretius—

> sensiferos motus quae didit prima per artus.
> prima cietur enim, parvis perfecta figuris. iii. 245-6.

> sic calor atque aer et venti caeca potestas
> mixta creant unam naturam et mobilis illa
> vis, initum motus ab se quae dividit ollis,
> sensifer unde oritur primum per viscera motus.
> iii. 269-72.

[2] The power of the soul, which is of so subtle a nature, thus to move the weight of our body, depends entirely on the close union of the two from the very first—

over the body: 'his own will makes for each man a beginning, and from this beginning the motions stream through the limbs.' Then he points out carefully how the body cannot move all at once, but 'the whole stock of matter over the whole body' must be sought out and stirred up *by degrees* till at length the frame obeys the bidding of the will. Over and over again he thus asserts the idea of a gradation of moving force, starting from the infinitely slight motion of the tiniest soul-atoms and growing gradually stronger and stronger till the body moves. This gradual increase of motion furnishes Lucretius with an argument for the existence of Free-will. When a man is pushed by some outer power, the body moves all at once, and continues moving till something within can stop it.[1] When we move of ourselves the motion is at first very slight, starting from the heart, and increases by degrees till the whole matter of the body is put in motion and we go whither we will. Thus —and most subtly—does his system link the mightiest efforts of the moving body with the Declination of the atoms, a motion so inexpressibly slight.

This view of Epicurean psychology enables us to see how subtle a part 'Declination' plays in Lucretius's system. The principle of atomic Declination was the most characteristic and, in several ways, one of the most important of Epicurean

> nonne vides etiam quam magno pondere nobis
> sustineat corpus tenuissima vis animai
> propterea quia tam coniuncta atque uniter apta est. v. 556-8.

The whole paragraph (534-63) illustrates the same notion. Again, at iv. 877-906, the animus is able to move the anima with ease, because the two are in close union,

> et facilest factu quoniam coniuncta tenetur.

Compare iii. 136:—

> animum atque animam dico coniuncta teneri
> inter se atque unam naturam conficere ex se.

[1] Coleridge uses an exactly similar illustration.

doctrines. One of its aims is to explain our Free-will. But for the fact that the atoms have the power at will to swerve slightly from their places 'at quite uncertain times and uncertain spots,' we should be the slaves of necessity. But for this power in the atoms, of which our souls are formed, we should not be able 'to move, nor yet to alter the direction of our motions not at a fixed time or fixed place, but just as our mind has prompted.' How, then, does Lucretius conceive Will to set the human body in motion? Our account of the four principles which compose the soul makes this plainer, and we can now see how subtly the doctrine of Declination is adapted to Epicurean psychology. The human Will acts directly on the finer atoms composing the fourth principle. 'Nothing else is formed of such small and smooth atoms' as this: *therefore nothing else can be so easily set in motion, or by so small a force.* The human Will (*i.e.*, the soul-atoms which possess volition) needs to exert but the smallest amount of force *at the outset* in order to produce the mightiest physical efforts of the body. We shall quote a passage from Sir John Herschel's paper on 'The Origin of Force,' one of the weightiest words of English speculative thought, which will help us to realize the principle here implied by Lucretius—

'The control we possess over the external world we are sure must arise from a capacity somehow inherent in the intellectual part of our nature, to originate or call into action this one and only agent which matter obeys in its changes of form and situation. . . . The actual *force* necessary to be *originated* to give rise to the utmost imaginable exertion of animal force in any case, may be no greater than is required to remove a single material molecule from its place through a space inconceivably minute—no more, in comparison with the dynamical force *disengaged*, directly or indirectly, by the act, than the pull of a hair-trigger in comparison with the force of the mine which it explodes. But without the power to make *some* material disposition, to originate *some* movement, or to change, at least temporarily, the amount of dynamical force appropriate to some one or more material molecules, the mechanical results of human or animal volition are inconceivable.[1] It matters not that we are ignorant of the mode in which this is performed. It suffices to bring the origination of dynamical power, to however small an extent, within the domain of acknowledged personality.'

Surely Lucretius (or his master Epicurus) must have foreseen that Will must, somehow or other, exert *force* on matter to make it move, also that its method of action might lie in giving a very slight initial impulse, from which by gradual increase the very greatest actual force might finally be developed. Lucretius's conception of Declination as a movement so exceedingly slight, the tiny soul-atom swerving from the straight line 'not more than the least possible' degree at the impulse of its own Free-will—does not this come pretty near to Herschel's 'no greater force than is required to remove a single material molecule from its place through a space inconceivably minute.' The principle involved is the same. It is not difficult to conceive Will acting on Matter in some such manner as this. Much in the same way, it is the arm of the engineer, deep down in the centre of the ship, moving a lever-handle a few inches, which actually sets in motion the huge Atlantic steamer weighing thousands of tons. Here, too, the beginning of motion was exceedingly small—the movement of the engineer's arm, allowing the steam to pass into the cylinder —but through a series of mechanical arrangements it results in an enormous force, which can hurl the sea-monster against wind and tide at the speed of three hundred miles a day. But we must return to this subject of Will and its relation to Force.[2]

[1] This sentence of Herschel's at once reminds us of the following words of Lucretius, 'If the first-beginnings do not swerve, and by so doing *produce some commencement of motion* so as to break through the decrees of Fate, that cause may not from everlasting follow cause; if it be not so, how do all living creatures upon the earth possess this power? (Free-will).'—Lucretius, ii. 251, ff.

[2] In his fourth book Lucretius uses our very illustration from machinery:—

Epicurus's notion of a fourth substance is, indeed, rather surprising. No doubt he thought by making matter as fine as possible to solve the difficulty of mind and matter, of the origin of Sensation—the side on which Materialism is as weak to-day as it was two thousand years ago.[1] To ancient

nec tamen illud in his rebus mirabile constat,
tantula quod tantum corpus corpuscula possunt
contorquere et onus totum convertere nostrum
* * * * *
multaque per trocleas et tympana *pondere magno*
commovet atque *levi* sustollit machina nisu.

In this passage (iv. 877-906) Lucretius again describes the process according to which will moves the human body. First of all, before the mind wills anything, an idol must strike it, *e.g.*, an idol of walking before we will to walk. The stages are as follows—

primum simulacra meandi
accidere atque animum pulsare . . . .
inde voluntas fit.

Next, when the mind has willed to walk, at once the mind (animus) strikes the soul (anima), which is scattered all over the body; next the anima strikes the various members of the body—

atque ita tota
paulatim moles protruditur atque movetur.

[1] Probably no one has stated this central difficulty of Materialism with more force, from one side at least, than Tyndall in his paper on 'Virchow and Evolution' ('Science Fragments,' vol. ii.). After computing material energy as developed in the brain by foot-pounds, he says, 'Consciousness on this view is a kind of by-product, inexpressible in terms of force and motion, and inessential to the molecular changes going on in the brain.' Again he inquires, 'What is the causal connection between molecular motions and states of consciousness?' 'My answer is,' he continues, 'I do not see the connection, nor am I acquainted with anybody who does. It is no explanation to say that the objective and subjective are two sides of one and the same phenomenon. Why should the phenomenon have two sides? This is the very core of the difficulty. There are plenty of molecular motions which do not exhibit this two-sidedness. Does water think or feel when it runs into frost-ferns upon a window-pane? If not, why should the molecular motion of the brain be yoked to this mysterious companion—consciousness? We can form a coherent picture of all the purely physical processes—the stirring of the brain, the thrilling of the

notions, matter thus refined must have come near enough to our definition of spirit. But the difficulty remains the same as ever. Given dead atoms, assorted according to coarseness or fineness into dead-matter atoms, organic-matter atoms, and soul-atoms, as minute as you like to make them, of any shape or moving in any order you like to conceive—how comes Sensation out of these? What or who has Sensation? Even supposing the individual atoms to be sentient (though Lucretius allows them Will alone), how comes it that so many isolated sensations of atoms separated by void are combined into one personal Consciousness?[1] Lucretius makes no attempt to answer this, except in stating that Sensation exists only in

nerves, the discharging of the muscles, and all the subsequent motions of the organism. We are here dealing with mechanical problems which are mentally presentable. But we can form no picture of the process whereby consciousness emerges, either as a necessary link or as an accidental by-product of this series of actions.' Therefore Tyndall asserts 'Between molecular mechanics (the interaction of the brain molecules) and consciousness is interposed a fissure over which the ladder of physical reasoning is incompetent to carry us.' Frankly and forcibly as Tyndall has stated this conclusion, what is the distinctive tone in his manner of approaching such questions which so often jars upon us, even when, as now, he helps us to realize them more keenly? An acute critic of Tyndall's (Mr. C. W. Richmond) has said, referring to the passage just quoted, that men of Tyndall's way of thinking 'recognize this chasm, this "fissure" which their ladder is too short to cross, under an illusion common in the case of those who limit their studies to physical nature. They place themselves in idea *on the wrong side of the gap*. They think they can approach the problems of mind from the side of matter and try in vain to lay the plank across.' This criticism applies pretty closely.—Picton makes a very ingenious attempt to realize the difficulty of explaining consciousness (of colour, for example) from any vibration of particles ('Mystery of Matter,' p. 35).

[1] According to the Epicurean doctrine of void—absolutely empty space between the atoms—no attempt can be offered at an explanation of this side of the difficulty. Might not the modern conception of ether as an omnipresent fluid, 'continuously filling space,' be pressed into service here? Such a fluid might be conceived to convey sensation between atom and atom. But, in order to be even vaguely conceivable by thought, would not this imply that the ether is sentient?

the Organism—in the animal body as a whole, soul and flesh together—not in its parts severally.[1]

Epicurus does not help us any more than Lucretius to understand the origin of Consciousness out of dead atoms, what we may call (to adopt a phrase of Schopenhauer's) 'the passage out of darkness into light.' According to Epicurus sensation is produced by the joint acting and common motions of body and soul. 'We must certainly admit,' he says, 'that it is the soul which is the chief cause of sensation. Yet it would not enjoy this power, if it were not environed by the rest of the body. . . . The soul transmits [sensation] to the body which is in union with it, and . . . . by a faculty of its own, it straightway realizes within itself the affection of sensation, and communicates it to the body in virtue of their close neighbourhood and sympathy.'[2] This reminds us of Schopenhauer's explanation of the same subject:—'An extremely small portion of this matter, which we call brain or ganglion according to its degree of organization or complexity, possesses the marvellous property of expressing in itself everything which acts upon it; *it is like a mirror in which the will is reflected and recognizes itself in all its degrees.*'[3] . . . 'This perceiving and conscious I is in relation to the Will [this including the notion of the outer world] what the image formed in the focus of a concave mirror is to the mirror itself,

[1] Bk. ii., ll. 904-14. Lange correctly remarks on this passage that 'this sensation of the whole' can in no case exist 'without the whole having also a certain Being.' 'This organic Whole is quite a new principle in addition to those of atoms and empty space.'

[2] Καὶ *μὴν καὶ ὅτι ἔχει ἡ ψύχη τῆς αἰσθήσεως τὴν πλείστην αἰτίαν δεῖ κατέχειν· οὐ μὴν εἰλήφει ἂν ταύτην, εἰ μὴ ὑπὸ τοῦ λοίπου ἀθροίσματος ἐστεγάζετό πως.* Compare the passage just following, where Epicurus says that we cannot conceive the soul to have sensation apart from the body, *μὴ ἐν τούτῳ τῷ συστήματι ταῖς κινήσεσι ταύταις χρώμενον, ὅταν τὰ στεγάζοντα καὶ περιέχοντα μὴ τοιαῦτα ᾖ ἐν οἷς νῦν οὖσα ἔχει ταύτας τάς κινήσεις.* (Diog. L., x. From the letter to Herodotus.)

[3] Ribot, 'La Philosophie de Schopenhauer,' pp. 158-9.

and like this image, it has only a conditional and apparent reality.'[1] However ingenious the illustration may be, and it is from Schopenhauer's standpoint an exceedingly apt one, it still explains nothing—any more than Epicurus's formula does. Both say simply, 'The soul has sensation because it has sensation,'—nothing more!

Lucretius is, indeed, singularly blind to all that is implied in Life. Most strange sometimes do his words sound; for instance, 'Never suppose that Life, the quality which we see floating on the surface of things, at one time born, then all at once perishing, can be inherent in the eternal atoms.'[2] A most strange conjunction of terms—*Eternal Atoms* but Life fleeting and merely superficial! Lucretius probably saw nothing jarring here, so sternly had he schooled his mind—doubtless after a period of questioning and struggle, fixedly put out of our sight in his poem—to accept the dogmas of his sect. He proceeds to give about thirty arguments for the mortality of the soul;[3] his reasoning, still striking, must have been more so in his own day, when assuredly it sounded new and scientific, and the very wording carried something impressive. Next he gathers all the strength of his poetry in one long torrent of eloquence to reconcile men to the thought of Eternal Death—not in any way going about to avoid the phrase—indeed, he uses the very words *mors immortalis, mors æterna,* with a most strange deadness of sense for what they imply. If at first he faces the dreadful fact with stern Roman determination, he seems at length to dwell on the thought of annihilation at death with indifference, sometimes almost with delight. It is true he shows no slightest power to conceive a happy life in the

[1] Ribot, 'La Philosophie de Schopenhauer,' p. 40.

[2] neve putes aeterna penes residere potesse
corpora prima quod in summis fluitare videmus
rebus et interdum nasci subitoque perire. ii. 1010-12.

Compare l. 1006 and the context.

[3] iii. 417-829.

other world. Yet, even while remembering that during his lifetime Rome was a slaughter-house, and all Italy bathed in blood, and, moreover, that priestly superstition painted the life to come as unjust and as horrible as the present was, even still there must have been *something abnormal, something unhealthy in the nature of Lucretius* before one great human instinct could have been thus seared within him.

CHAPTER VII.

ATOMIC DECLINATION AND FREE-WILL.

THE reader who has thus far patiently followed us along the intricate passages of Lucretius's atomic materialism, has now reached a point where the poet's path appears most tortuous and darkest. Here Lucretius seems, as it were, to descend underground and to lead us nowhere at all, except persistently away from the sunlight. Yet even here, where he might with reason seem to have completely lost his track, we shall catch a vivid gleam of light far away.

Epicurus, as we before explained, asserts that the atoms have the power of Free-will and are able 'to decline,' that is, to swerve very slightly [1] from their path 'at no fixed time or place,' but according to their own volition. No doubt this doctrine is remarkable enough. It seemed very absurd indeed to the philosophers of Greece and Rome, the contemporary and later critics of Epicurus, who never tire of laughing at it [2]—but it does not appear so ridiculous to philosophical students at present. In particular, M. Guyau, probably the most acute

[1] Nec plus quam minimum. ii. 244.

paulum
tantum quod momen mutatum dicere possis. ii. 219-20.

perpaulum quo nihil posset fieri minus (Cicero, de fin., 19); ἐλάχιστον ('De Fato,' 9); ἄτομον παρεγκλῖναι μίαν ἐπὶ τοὐλάχιστον (Plutarch, 'De Solertia Anim.,' 7); ἀκαρές ἐγκλῖναι ('De anim. procr.,' 6).

[2] See Cicero and Plutarch, *passim*.

critic whom Epicurus has ever met with,[1] finds it, as we shall see, thoroughly coherent and logical. It is worth trying to understand by what arguments so shrewd a thinker as Epicurus could justify so audacious a notion.

In this atomic swerving Epicurus had a twofold purpose. Firstly, the mere fact of declination is enough to bring his atoms into contact. In the second place, it will be noticed that he carefully qualifies the declination ; it takes place

> Incerto tempore ferme
> incertisque locis,

'at quite uncertain times and uncertain spots.' He has a reason for so doing. This is solely for the purpose of solving a great philosophical question, over which controversy raged as fiercely in Lucretius's day as it has done in our own—the question of Free-will as opposed to Fate and Necessity. Epicurus emphatically maintained the doctrine of Free-will in opposition to Heraclitus, Democritus, and most of the Stoics, who held an everlasting and inexorable Necessity, and denied the existence of individual self-will. By his conception of a regular order in Nature, Epicurus had delivered himself from the notion of Gods constantly interfering with the world and filling men with the fear of unseen adverse powers. But, when escaped from this, he found himself all at once confronted by the yet more terrible phantom of Necessity—destroying Free-will and transforming man into the slave-machine of Fate. How did he again escape from so dreadful an alternative? It was by an expedient singularly bold and simple.[2]

[1] In addition to his able work on Epicurus, M. Guyau is very favourably known in this country, as a critic of philosophical systems, by his work, 'La Morale anglaise contemporaine,' in which he discusses modern British philosophy from Bentham to Spencer and Darwin.

[2] We may refer here to our later chapter on M. Guyau's book. In some brilliant pages from which we have there quoted, M. Guyau has realized, more vividly than any other writer, the dilemma which con-

The passage in which Lucretius defends Declination is one of the most interesting in his poem. He is fighting at close quarters for the most characteristic and distinctive doctrine of his own and his master's creed. These forty lines must be taken as the product of a philosophical controversy, as fierce and voluminous as any that have raged in our own day, or been discussed in the pages of our philosophical reviews. We hear the clash of the combatants' weapons as we read this short notice, containing a phrase or two of decidedly controversial coinage. The whole paragraph[1] (which we now quote in full) is most closely reasoned. No word is thrown away.

fronted Epicurus, and the unique solution of it which he attempted. So far as this, we shall find M. Guyau an admirable guide.

[1] Denique si semper motus conectitur omnis
et vetere exoritur semper novus ordine certo,
nec declinando faciunt primordia motus
principium quoddam quod fati foedera rumpat,
ex infinito ne causam causa sequatur,
libera per terras unde haec animantibus exstat
unde est haec, inquam, fatis avolsa potestas,
per quam progredimur quo ducit quemque voluntas,
declinamus item motus nec tempore certo
nec regione loci certa, sed ubi ipsa tulit mens?
nam dubio procul his rebus sua cuique voluntas
principium dat et hinc motus per membra rigantur.
nonne vides etiam, patefactis tempore puncto,
carceribus, non posse tamen prorumpere equorum
vim cupidam tam de subito quam mens avet ipsa?
omnis enim totum per corpus materiai
copia conquiri debet, concita per artus
omnis ut studium mentis conixa sequatur;
ut videas initum motus a corde creari
ex animique voluntate id procedere primum,
inde dari porro per totum corpus et artus.
nec similest ut cum impulsi procedimus ictu
viribus alterius magnis magnoque coactu:
nam tum materiem totius corporis omnem
perspicuumst nobis invitis ire rapique,
donec eam refrenavit per membra voluntas.
iamne vides igitur, quamquam vis extera multos

'For beyond a doubt, in these things, for each man his own will makes the beginning, and from this the motions are spread in a stream through the limbs. Do you not see too, when the barrier has at a given moment been flung open, that still the eager might of the horses is not able to break away so suddenly as their mind longs to do? For the whole store of matter throughout the whole body must be sought out in order that it may be stirred up all over the frame, and may with one united effort obey the desire of the mind, so that you see that the beginning of motion is born from the heart, and that it issues in the first place from the desire of the mind. From this it is next distributed through the whole of the body and limbs. It is quite different when we move on impelled by a blow, inflicted by the strong power and strong compulsion of another,[1] for then it is plain that all the matter of the whole body moves and is hurried on against our desire until the will has reined it in throughout the limbs. Do you see now, therefore, although an outside force in many cases drives men on and often compels them to go forward against their will and to be hurried on headlong, that still there is within our breast something which is able to fight against and to resist it? And at the beck of this something also, the whole amount of matter is compelled sometimes to change its course all over the limbs and frame, and after it has been driven forward, it is

pellat, et invitos cogat procedere saepe
praecipitesque rapi, tamen esse in pectore nostro
quiddam quod contra pugnare obstareque possit?
cuius ad arbitrium quoque copia materiai
cogitur interdum flecti per membra, per artus,
et proiecta refrenatur retroque residit.
quare in seminibus quoque idem fateare necessest,
esse aliam praeter plagas et pondera causam
motibus, unde haec est nobis innata potestas,
de nilo quoniam fieri nil posse videmus.
pondus enim prohibet ne plagis omnia fiant
externa quasi vi; sed ne mens ipsa necessum
intestinum habeat cunctis in rebus agendis
et devicta quasi hoc cogatur ferre patique,
id facit exiguum clinamen principiorum
nec regione loci certa nec tempore certo. ii. 251-93.

1 Viribus alterius magnis magnoque coactu.

The repetition of *magnis* in this line has a pointed force: it is intended to contrast with the small beginning of self-originated motion.

reined in and settles back into its place. *Wherefore in atoms as well*[1] *you must admit the same, that in addition to blows*[2] *and weights there is another cause of motions whence this power has been begotten in us, since we see that nothing can come from nothing.* For weight forbids that things should be done altogether by the agency of blows, through as it were an outward force ; but that the mind itself does not feel an internal necessity in all its actions, and is not completely overpowered, as it were, and compelled to bear and put up with this, is caused by a minute swerving of the first-beginnings at no fixed part of space and at no fixed time.'

It is worth while to retrace Lucretius's reasoning, and keep as close to it as we can. It is pretty nearly this.

If the whole world of nature and man is a mechanism in which cause follows cause and motion follows motion in a fixed order from everlasting, we could not possibly have Free-will. Here Lucretius appeals emphatically to our inner consciousness. 'Beyond a doubt' we feel and know that, when we move, our own will makes a beginning of motion, and this motion gradually increases. We see the same thing in horses which, when the signal has been given in a race, cannot start at once, though eager to do so. The motion has to begin at the heart, and spread through the limbs. Lucretius considers the gradualness felt and discerned in the movements of living things (which originate from within and gradually grow) as a proof of Free-will. Movement caused by Necessity is quite different. The tendency of outside force is to make the body acted on move all at once. Thus when a man is violently pushed, he moves all at once. Here again (ll. 271-83) Lucretius makes a confident appeal to the facts of personal experience,—a consciousness so clear and vivid that it cannot be countersaid.[3] How clearly we discriminate between moving of our

[1] *i.e.*, in atoms as well as in human beings.

[2] plagae, *i.e.*, the blows of atoms in collision. *Plagae* is Lucretius's term for mechanical force.

[3] Mr. Henry Sidgwick, in his 'Method of Ethics,' has emphatically expressed the convincing force of this argument from individual experi-

own accord and being pushed! In the latter case we are propelled forward, but all the time we feel with positive conviction that there is 'something within our breast' which can resist the outer impulse, and finally checks it. The same is the case with atoms, too.[1] In the world of nature [2] all motion is caused by 'blows' and 'weights;' but there must be in all atoms, and particularly in the atoms forming the soul, another cause of movement, namely, the power to decline at will. This is necessary, since if the soul-atoms have not this power, while the soul has it, we violate the first principle—nothing can come from nothing. In thus applying his axiom 'ex nihilo nihil fit' (l. 287), Lucretius draws the final conclusion—a bold enough one, too—of a close-reasoned argument. He reasons thus: '*I cannot account for Free-will appearing in human beings, the highest product of atomic evolution, unless it were there from the first. If men have Free-will, then Matter which they come from must have Free-will too, since nothing can come from nothing.*'

Lucretius goes on, still speaking strictly of the mind, to discriminate pointedly between two kinds of necessity which may enslave it, an external one and an internal one—*externa vis* on the one hand, and *necessum intestinum* on the other.[3] The

ence. After first allowing that there is 'an almost overwhelming cumulative proof' that human Free-will does not exist, he speaks as follows: This evidence 'is more than balanced by a single argument on the other side, the immediate affirmation of consciousness in the moment of deliberate volition. . . . No amount of experience of the sway of motives ever tends to make me distrust my intuitive consciousness that in resolving, after deliberation, I exercise free choice as to which of the motives acting upon me shall prevail.'

[1] The inference conveyed by *quoque* must not be forgotten. It refers to the preceding illustrations of free-will action in men and animals. It means 'in atoms as well as in human beings.'

[2] In the appendix we have discussed an apparent difficulty in this part of Lucretius's argument.

[3] M. Guyau has clearly enforced this, and has very aptly shown with reference to this passage that ancient philosophers held weight (as causing

natural gravity of the mind-atoms would offer a certain amount[1] of resistance to 'blows' from without—that is, to the Necessity which reigns everywhere in Nature. But though weight is less of an externa vis, still it is fatal within the mind, and would, even more hopelessly, imprison mens ipsa—the mind within her own house.[2]

Therefore, if the mind is to be free, the atoms composing it must possess Free-will. Nothing else can deliver the mind from Necessity.

Such is the reasoning by which Lucretius proves that this marvellous power of Declination exists everywhere around us, in the atoms composing every form of Matter. What then comes of it? we cannot help asking. Does a power like this exist, yet produce no effect in the world? On this subject M. Guyau has built up a theory which he endeavours to prove from historical evidence to have been the actual teaching of Epicurus; nay more, M. Guyau seriously accepts this theory as scientifically true from a nineteenth century standpoint. According to him 'Declination' does not disappear without

motion to originate from within) to be less of the nature of necessity than an outside force was. Compare Cicero, 'De Fato,' xi., 'De ipsa atomo dici potest enim, quum per inane moveatur gravitate et pondere, sine causa moveri, quia nulla causa accedat extrinsecus.'

[1] These, it must be remembered, are exceedingly minute as compared with dead-matter atoms. The word omnia here is emphatic.

[2] M. Guyau (pp. 85-7) has done service by referring to the very ingenious attempt of Carneades to show that Declination was unnecessary because, just as it is the 'nature' of the atom to move by its own weight, similarly it is the 'nature' of the mind to move of its own accord. 'Thus,' says M. Guyau, 'by the idea of *nature*—that is to say, of a cause which should in reality be neither free nor necessary—Carneades hopes to reconcile the regularity of movements in the universe with their arbitrary freedom in man.' The 'nature' by which Carneades explained the difficulty appears to be a kind of 'inherent potency,' like that of which we have heard a good deal lately. It is not impossible that, as Guyau seems to imply, in this passage Lucretius is expressly combating the very arguments of Carneades.

any result, but produces wondrous effects in Nature. His theory is one of the most remarkable castles-in-the-air which have ever been raised, and will certainly not be forgotten in the history of philosophy.

In insisting on the importance of our personal consciousness of Free-will, and in reasoning back, after a fashion, from it to the origin of all Force, as well as in supposing Will to exist everywhere in every form of matter, Epicurus strongly reminds us of a very remarkable modern philosopher, Schopenhauer. According to Schopenhauer, the only direct knowledge which we have is that of our Will, and our personal consciousness of Will is the only key to the essence of nature and of natural forces. Science, he says, can never teach us what a law *is*, nor yet what a force *is*, nor what a cause *is*. All the physical sciences have one common defect—their method is exterior. They remind us of 'a man who wanders round and round a castle, vainly seeking for an entrance, and meanwhile sketches the outside of it.' It is only our own sense of free-will action, in the body by which we are rooted in the world, that enables us to solve the riddle, and tell what the essence of force is. Schopenhauer insists on reducing all the forces of nature to Will. 'I consider,' he says, 'every natural force as a Will. Will is essentially identical with all the forces which act in nature, the various manifestations of which belong to the species of which Will is the genus. *It is the direct consciousness which we have of Will which alone conducts us to the indirect knowledge of the other Forces.*' . . . . 'If, then, we reduce the notion of Force to that of Will, we reduce the unknown to a thing much better known, to the only thing directly known, and this greatly extends our knowledge. If, on the opposite, we reduce, as has hitherto been done, the concept of Will to that of Force, we abandon the only immediate knowledge of the world which we have: we allow it to be lost in an abstract conception which is derived from phenomena, and with which we shall never be able to get beyond them.' Many critics have charged Schopenhauer with simply exalting Force under the name of Will.[1] According to them, his 'Will' simply requires to be translated by the name of 'Force.' This criticism is only very partially true. We must remember that Schopenhauer's derivation of his doctrine from our own direct consciousness of effort, when we will any act, affords a real basis for his conception, and to a considerable extent justifies his application of the term Will to denote the essence of all things. In thus claiming that we understand the operation of natural forces only in virtue of our own acts of conscious Will, Schopenhauer can reasonably claim that he proceeds from the known to the unknown. If on one side, as he says, his 'Will' is 'only an x, an absolutely unknown quantity,' on the other it is, he asserts, 'infinitely better known to us, and more certain than all else.' In reality Schopenhauer's philosophy, which reduces the world to a 'manifestation of Will,' is a kind of Pantheism, but Pantheism of a kind peculiar to himself. We cannot here reproduce the elaborate argument in which Schopenhauer insists on reducing the concept of Force to that of Will.[2] There are several points of contact, some of them pretty close, between Schopenhauer and Lucretius, but at present we can only remind our readers how Lucretius too intimately associates Will with the origin of Force.

There are two aspects from which we may consider Lucretius's doctrine of atomic Declination.

I. In the first place, it is well to observe how simple, yet how

[1] So, according to M. Ribot ('La Philosophie de Schopenhauer,' Paris, 1874, p. 152), 'This blind unconscious Will can only be Force;' and again, p. 177, Schopenhauer 'places Force at the crown of things under the name of Will.' Schopenhauer himself utterly refuses to allow his concept of 'Will' to be thus reduced to Force. See M. Ribot, pp. 67-8.

[2] We may refer to the admirable chapter, 'La Volonté,' in Ribot's excellent little work above mentioned.

efficacious a part it plays in the Epicurean system. Its rôle is a double one. According to Lucretius the atoms, which are conceived to fall straight downwards, from all eternity, in parallel lines like drops of rain, would never meet and by meeting create the world—but that occasionally they swerve a very little from their path, and so come into collision. If we can move at will, Lucretius argues, therefore the atom must be able to move at will, and if it move at Free-will, it must go off the perpendicular line of Necessity, and by thus declining the atoms are brought into contact.—Thus the Epicurean explanation of Free-will at the same time explains the origin of the world.

II. The doctrine is far more interesting, however, when we consider the philosophical consequences implied in it.

1. The reasoning of Lucretius which we have before abstracted is cogent enough. We see that Lucretius is not merely daring but also logical in assigning Will to matter. That which is in the effect must also be in the cause; therefore, if Free-will be in man, it must also exist in the atoms of which he is composed. A human creature endowed with Free-will cannot come from atoms which do not possess volition. From Lucretius's standpoint, this is a logical Necessity. (It is certainly surprising, however, that the author of a materialistic system should have allowed the existence of Free-will at all.) Very noteworthy too is the manner in which Lucretius strengthens his argument by appealing to our personal consciousness of effort. The vividness and directness of this feeling is, he implies, something unanswerable. No arguments for Necessity can do away with it.

A theory substantially the same as Lucretius's, but more subtle, has been stated in our own time. We refer to Professor Clifford's doctrine of 'Mind-Stuff,' which runs so closely parallel with that of atomic Declination that it is worth while to give some account of it here. The two doctrines illuminate each other. Professor Clifford's theory, stated shortly, is, as nearly as we can reproduce it, the following[1]:—

Along with every fact of consciousness in our mind, there goes some disturbance of nerve-matter. When a man is conscious of anything, 'there is something outside of him which is matter in motion, and that which corresponds inside of him is also matter in motion.' Both are made of the same stuff; the object outside and the optic ganglion are both matter, and that matter is made of molecules moving about in ether. Thus, whenever the ganglion of the brain is disturbed because certain pieces of grey matter there have arranged themselves in the figure of a square, the consciousness of a square is produced in my mind. Thus there are two classes of facts which always run parallel, 'physical facts and mental facts.' But there exist far lower and less complex forms of feeling than such as make up human consciousness. 'We are obliged to assume, in order to save continuity in our belief, that along with every motion of matter, whether organic or inorganic, there is some fact which corresponds to the mental fact in ourselves. The mental fact in ourselves is an exceedingly complex thing; so also our brain is an exceedingly complex thing. We may assume that the quasi-mental fact which corresponds and which goes along with the motion of every particle of matter, is of such inconceivable simplicity, as compared with our own mental fact, with our consciousness, as the motion of a molecule of matter is of inconceivable simplicity when compared with the motion in our brain.' According to Professor Clifford our consciousness is a very complex thing indeed. 'When a stream of feelings is so compact together that at each instant it consists of (1) new feelings, (2) fainter repetitions of previous ones, and (3) links

[1] It is worked out in two papers, 'Body and Mind' and 'On the nature of Things-in-themselves,' in his 'Essays and Remains,' vol. ii., 1879. The first statement of the doctrine is said to be found in Wundt's 'Grundzüge der physiologischen Psychologie.'

connecting these repetitions, the stream is called a consciousness.' No single feeling of ours is a unit. Every feeling of mine is a most complex structure, built up from a great many different elementary feelings which are grouped together in various ways, just as the action of my brain is made up of a great many elementary actions in different parts of it, grouped together in the same ways. Thus each elementary feeling corresponds to a special, comparatively simple, change of nerve-matter. It is a popular error to suppose that a feeling cannot exist by itself without forming part of some consciousness. If then we go back along the line from the organic to the inorganic, and if 'according to the complexity of the organism is the complexity of the consciousness,' where are we to stop? Where does the breach of continuity take place? Where does some degree of feeling cease to accompany the motion of matter? 'There is only one way out of the difficulty, and to that we are driven. Consciousness is a complex of ejective facts,—of elementary feelings, or rather of those remoter elements which cannot even be felt, but of which the simplest feeling is built up. Such elementary ejective facts go along with the action of every organism, however simple;' but it is only with a complex nervous structure that the complication of feelings called Consciousness co-exists. 'But as the line of ascent is unbroken, and must end at last in inorganic matter, we have no choice but to admit that *every motion of matter is simultaneous with some ejective fact or event* [*i.e.*, elementary feeling] *which might be part of a consciousness*.' From this follows the important corollary, 'A feeling can exist by itself without forming part of a consciousness. It does not depend for its existence on the consciousness of which it may form a part.' These elements of feeling, of which our simplest ordinary feeling is a complex, Professor Clifford calls *Mind-Stuff*. 'A moving molecule of inorganic matter does not possess mind or consciousness, but it possesses a small piece of mind-stuff.'

By a process of reasoning which is very hard to apprehend,[1] Professor Clifford decides that the thing-in-itself, the reality which underlies all matter, is the same stuff which, when compounded together, produces mind. It is not the same thing as mind, but is made of the same stuff. Thus the universe consists entirely of mind-stuff. 'There is no matter without something like mind behind it.' According to this theory all matter is mind-stuff *on one side,* on what may be called its interior side. We naturally ask then, Is there no mind-stuff on a vast scale, answering to the inner side of the vast material universe? and may we not then conceive the reality underlying the whole world as one great mind? No, says Professor Clifford, 'Human consciousness is, so far as we know, the highest that exists.'

Of course Professor Clifford's theory does not by any means explain the origin of Consciousness. 'Every mental picture,' says Clifford, 'is made up of exceedingly simple mental facts, so simple that I feel them only in groups.' For one thing, if a single elementary feeling does not produce consciousness, why should mere complexity produce it? If every molecule of my body possesses some degree of sentiency, does this account for my conviction of personality? Yet the theory, however insufficient, is instructive. The materialist feels that it is a hopeless task to explain the origin of Consciousness or of Free-will out of dead atoms, without some break in the continuity of development, some new entrance of Energy into the field. His only escape from the difficulty is this:—the atoms are not 'utterly dead,' but contain in a faint and weak form the faculties of consciousness and mind which are found in the highest product of Evolution, man. Thus Professor Clifford, in order to explain the evolution of Mind from atoms, asserts that every atom of

[1] 'Lectures and Essays,' vol. ii., pp. 85-7.

matter corresponds to an atom of Mind-Stuff, that is, of something analogous to Mind. He thus builds up Mind out of a multitude of mind-atoms, that is to say, of elementary feelings which can exist by themselves as 'individuals,' *simplicitate*, as much as can the Lucretian atoms, but which are almost as small in comparison with the consciousness of any one human being as Lucretius's atoms are in comparison with a human body. Lucretius again, who believes in Free-will, can only explain it by assigning Free-will to the atoms. The reasoning of both, starting from a similar standpoint, is substantially the same, and the two theories of 'Mind-Stuff' and of Atomic Declination deserve to be placed side by side. Both are based upon the same principle,

> unde haec est nobis innata potestas
> de nilo quoniam fieri nil posse videmus,

and apply it with equal boldness. Both moreover show to us Materialism confessing its own weakness to account, unaided, for the origin of Mind.

2. Lucretius is a most ardent believer in individual Free-will, which indeed was one main dogma of Epicureanism. He supposes Declination to be amply potent in the way of making this possible. In giving an account of Epicurean psychology, we quoted a most important passage from Herschel, showing that unless we suppose the will to exert some amount of Force, however slight, we can hardly conceive its action. Is not this same principle most clearly implied in Lucretius's account, already explained at full, of the method by which Will sets our body in motion? The Will acts directly on the smallest soul-atoms, those of the fourth principle; these in turn move the coarser and more coarse till the whole body is set in motion, but this results entirely from a very slight original impulse given by Will. However tiny and easily moved these soul-atoms are, it still requires some exercise of Force, however slight, to set them in motion. Lucretius's explanation of the

manner in which Will makes the body move, *most distinctly implies a power in the Will to originate Force.*

But what is to be thought of the other use of Declination, namely, in order to bring the falling atoms into collision and render them a source of Power? But for Declination, Lucretius believes, these atoms would 'fall downwards through the void, like drops of rain,' barrenly, and 'Nature would never have produced anything.' It alone brings them into collision so that they become a source of Power,—the source from which all the Energy of the universe is derived. Can it be said that, in this notion, Lucretius is carrying out the idea of Personal Causation in the same way as Martineau and Herschel have done? Of course Lucretius's materialism could never have allowed him to rise to the thought that *all* Force may be the outcome of Will—'that any causal power other than Will is unthinkable.' But if not fully, did Lucretius grasp this thought at all?

Surprising as the notion may seem, we still venture to think that Lucretius may have been led from his strong belief in individual Free-will to this conception of Will residing in the atoms and bringing them into creative action before the world was made. Lucretius's language in discussing Free-will shows a *strong consciousness* of Personal Causation, as exerted by man himself. Indeed Lucretius refers almost triumphantly, as a convincing argument for the existence of Free-will, to our vivid personal consciousness of exerting effort when we will to move. 'Beyond a doubt,' he says, 'his own will makes a beginning for each;' and he goes on to describe the process. Now is not the thought natural enough? If Free-will originates action in men (and Lucretius grasps this truth most strongly), may it not, in the world of Matter too, give to the atoms the first impulse which finally results in creation. This is all that Lucretius supposes Declination to do. *It gives the initial impulse* in both cases. As motion in the human body must be begun by the Will, so Will only can be supposed to give the first impulse, without which the atoms could never create. We have seen that the step from Personal Causation back to first-existing Causation is natural enough. We believe that Lucretius too may have had a perception of this in a blind unreasoning way,—partly, it would seem, as a corollary from his belief in Will originating bodily motion.

3. At first sight, it is true, the volition of the dead atoms is hardly what we call Will. On this point again Schopenhauer comes into remarkable agreement with Epicurus. According to Schopenhauer too, Will is essentially unconscious and becomes conscious only by accident;—it exists everywhere as a blind tendency in Matter, but mounts up by degrees through vegetable and animal life till it appears in man, in its highest degree of objectivation, as intelligent and distinctly conscious of itself. It is remarkable that Epicurus and Schopenhauer *both conceive Will to be essentially unconscious* and to exist everywhere in Matter as a blind force, but, by some miraculous transformation which neither of them takes into account, to emerge in man as a conscious one.

Acts of volition on the part of dead atoms are, in truth, hardly thinkable. A dead Will is indeed an absurdity, but Will without life was one of the Epicurean dogmas which Lucretius, as a convert, had felt himself bound to swallow. He will not even admit to himself that there is anything unnatural in it. As we have said, he most emphatically insists that Matter is 'utterly dead.' After having reduced the whole universe to atoms, he makes merry, with somewhat ghastly laughter, over the idea of these little particles having life. If so, you may think of them 'laughing and weeping, and learnedly discussing Epicurean doctrines.'[1] To do this, and conceive them living is 'raving madness,' he says. Yet surely that is not more absurd than his own conception of atoms that have

[1] ii. 973-90.

wills of their own and can move to the right or left, 'as they wish,' *sponte sua,* but which are at the same time 'utterly dead.' It is for this reason, because his atoms act as if they were alive, and because a dead Will is an absurdity, that we formerly said that Lucretius *virtually* conceives Matter as living, and that really his theory of Matter involves—certainly in a very low form, perhaps in the very lowest form possible—Pantheism.

This statement may seem somewhat startling, especially when we consider what Lucretius's creed was—how thoroughgoing in many ways was his Materialism. But, logically, his position does amount to this. And, if it be true that Lucretius attributes Will to Matter, and, even in a blind and indirect way, assigns Will as the source of energy, does not this to some extent alter his position among the philosophic systems? Though at the cost of utter inconsistency, has not even he been *compelled* to admit within his scheme of evolution something besides dead Matter? For it must not be forgotten that the Matter which, he says, is so potent to evolve life, is Matter + Free-will.

It is, we hope, now evident that the strange doctrine of Free-will residing in the falling atoms before the beginnings of the world and existing everywhere around us in every molecule of Matter, is not so entirely arbitrary as at first it seemed. Underlying it there is an element of truth.

Note on Professor Clifford's Theory of Mind-Stuff as anticipated by Gassendi.

It is worth while to point out that Professor Clifford's very notable theory was substantially anticipated by an ardent modern disciple of Epicurus, who published his great work on

Epicurean science and ethics two hundred years ago, Peter Gassendi. 'As of a certainty,' says Gassendi, 'whatever is in the fire existed first in the wood, but in a latent state, . . . thus whatever substance is in the soul that exists and moves within the body of an animal, the same first existed in the food, seed, or other matter which produced it.' Turning to the lower forms of life, he recognizes in plants[1] in a weak form Hunger and Appetite as seen in their assimilation of food, Taste and Touch as seen in their preference of one soil as nutriment to another, and Pleasure as shown in their joyful outburst of growth in spring after the winter sleep. In plants he finds also traces of Habit. Even in inanimate matter he sees something like Consciousness, and notably so in the magnet. In a very remarkable passage[2] he shows how the iron is impelled to the magnet 'by a kind of desire' (*appetitu quodam*). The magnet 'appears to turn towards itself and to attract the very soul, as it were, of the iron, which by its own force carries with it to the magnet the whole mass of the iron,' just as our own

[1] 'Epicuri Philosophia,' 3rd edition, Leyden, 1675, vol. i., pp. 263-4.

[2] Vol. i., pp. 202-3. 'There appears,' says Gassendi, 'to be in the magnet and in the iron a certain quality analogous to Consciousness. . . . For as an animal when struck by the aspect of an external object, immediately desires it, and is impelled towards it, thus the lesser magnet or the iron, as soon as it is smitten by the form of a greater or more powerful magnet, is impelled towards it by a kind of desire (appetitu quodam rapitur ad ipsum).' As with objects which men see, so with the magnet—its attraction can only work subject to the condition of distance. 'Just as a sensible object attracts to it the mind of man, and the mind moved by desire approaches his body to that object, even so the magnet also, through its transfused form, appears to turn towards itself and to attract the very soul, as it were (or, as it were, the flower of the substance), of the iron, which by its own force carries with it towards the magnet the whole mass of the iron. It would be difficult to believe how a thing so slight as is our conscious soul, . . . can move the weight of our body, which is so heavy and inert, unless experience assured us of it. What else, then, should we believe to be in the iron except a soul, or at least something analogous to soul, which, although very slight, still is able to move the rest of the mass, very heavy and inert though it be.'

soul, though light, moves our body. There must be in the iron 'a soul, or at least something analogous to soul.' Thus in the magnet as well as in plants Gassendi finds 'a foreshadowing of consciousness' (*adumbratio sensus*), and he traces this gradually strengthening as it mounts up through shell-fish and worms to the higher animals. 'Nature,' he says, 'is not accustomed to pass from one extreme to another except through intermediate stages. Thus, for example, the fruits of trees become sweet instead of bitter, fragrant from scentless, yellow from green, by a gradation so imperceptible, that at the beginning nothing is discerned of that quality which is to be, and at the end generally nothing of that which was at the beginning, so that we may thus understand that *unconscious matter becomes conscious by an exactly similar gradation, which certainly it is not within human power to trace*'[1] (vol. i., p. 270). Again, because all living things, even the meanest, those spontaneously generated, come from seminal molecules, each after its kind, which have existed either from the beginning of the world or from a later time, 'for this reason it cannot absolutely be said that conscious things come from non-conscious, but rather from particles which, though they do not actually possess consciousness, nevertheless actually are or do contain the elements of consciousness (*principia sensus*).' Are not these 'elements of consciousness' contained in Gassendi's molecules much the same as Clifford's simple elementary feelings or Mind-Stuff? Gassendi does not, it is true, say that every separate atom contains an element of sensation. In reality, by his distinction (vol. i., p. 272) between prima materies or non-conscious atoms, and secunda materies or molecules which possess in a faint form the rudiments of sensation, he does not at all escape the difficulty of the origin of consciousness, which indeed he, like Epicurus, very slightly realizes. At the same time his sugges-

[1] He even defends this opinion by the aid of Lucr., ii. 931-43, which he totally misinterprets.

tion, so cautiously expressed,[1] as if he realized its possible bearing as a materialistic argument, of a gradual, imperceptible progress, from molecules containing 'the rudiments of sensation' up to complete living consciousness, shows the acuteness of Gassendi. This theory of Gassendi's and Guyau's doctrine of 'Spontaneity' are both the outcome of a revival of Epicurus's scientific system in the sixteenth and nineteenth centuries respectively. The tree of Epicureanism, which has even in modern times twice broken out into fresh bud and leaf, and put forth on French soil two such remarkable offshoots, cannot even yet be called dead.

[1] 'Propono seu potius balbutiens attingo, ut quatenus licet, insinuem progressum quo res videntur evadere ex insensilibus sensiles' (vol. i., p. 270).

CHAPTER VIII.

HOW LUCRETIUS CONCEIVES THE WORLD;—HOW DOES IT HOLD ITS PLACE IN THE UNIVERSE?

IT is somewhat difficult to realize how Lucretius conceives the world. Without drawing from his words any unwarranted inference, we wish here to point out that the world as described by Lucretius fulfils most of the chief conditions of an organism, as these are laid down by himself.

In proof of this it is worth while to state fully his account of the world's gradual growth, maturity, and present decay. At the end of the Second Book he shows that the world was not always of the same size as now,[1] but has gradually increased. The earth and sea have grown greater, and the vault of heaven higher, according as each has been fed by matter from without. 'Since the birth-time of the world, and the first day of being of earth and sea, and the first forming of the sun, many bodies have been added from without, many seeds have been added all around, which the great universe has contributed in its tossing to and fro,[2] that from these bodies the sea and lands might increase, and from them the house of heaven might enlarge its room and raise its high dome far above earth, and that air might rise up also. For all bodies from all quarters are distributed by blows, to each thing the atoms of its own kind, and all depart to their own classes; moisture joins moisture, earth increases from an earthy base, and fires forge fires, and

[1] ii. 1105-74.

[2] Mr. Munro thus translates iaculando.

ether ether.'[1] In the course of time the world reached its utmost limit of growth and remained stationary, so long as the fresh matter absorbed by its 'life-arteries'[2] was no more than the amount of waste matter. This is nature's law of growth. It is thus that healthy and happy young creatures increase in stature while their bodies absorb more than they lose. 'Whenever the growth of anything has stopped, the larger and the wider it is, the more particles does it scatter in every direction and lose from itself, nor is food readily transmitted into all its arteries, and the amount is not sufficient to enable so large a quantity to rise up and be supplied in proportion to the copiousness of the drains from it.'[3] Food will now no

[1] Multaque post mundi tempus genitale diemque
primigenum maris et terrae solisque coortum
addita corpora sunt extrinsecus, addita circum
semina quae magnum iaculando contulit omne;
unde mare et terrae possent augescere et unde
appareret spatium coeli domus altaque tecta
tolleret a terris procul, et consurgeret aer.
nam sua cuique locis ex omnibus omnia plagis
corpora distribuuntur et ad sua saecla recedunt.
umor ad umorem, terreno corpore terra
crescit et ignem ignes procudunt aetheraque aether,
donique ad extremam crescendi perfica finem
omnia perduxit rerum natura creatrix. ii. 1105-17.

The constant atomic collisions have the effect of distributing this fresh matter, and enabling the atoms of each kind to join those similar to themselves, whether those of water, earth, fire, or ether. Munro thus explains the force of *plagis*: 'these blows of the atoms are, as we have so often seen before, the chief cause of the formation and conservation of things by enabling the atoms to clash and try all kinds of union until some suitable one is found.'

[2] Thus Mr. Munro translates:

ut fit ubi nilo iam plus est quod datur intra
vitalis venas quam quod fluit atque recedit.

[3] quippe etenim quanto est res amplior, augmine adempto,
et quo latior est, in cunctas undique partis
plura modo dispargit et ab se corpora mittit,
nec facile in venas cibus omnis diditur ei

longer sustain it, 'since the arteries refuse to hold what is sufficient, and nature does not furnish the amount needful.'[1] Thus when the world's waste comes to be greater than its fresh supplies of matter, it must ere long give way to the destructive forces ever at work, seeing that its substance is weakened by waste and unable to resist the battering of the torrent of atoms which are eternally striking it, and which it is now unable to assimilate into its substance.[2] 'In this way then the walls of the great world all round it shall be stormed and fall into decay and crumbling ruins.' This process of decay, Lucretius believes, has in his own day been long at work. The earth is now growing old and feeble, he says, and is able to bear nothing but small living creatures, such as worms or insects, the earth who once brought forth the huge bodies of wild beasts. Corn crops, vineyards, fruit and pastures once produced by the earth spontaneously and in abundance, now scarcely repay our utter toil. The fields are niggard of their produce, and the vines are small and exhausted.[3]

Professor Sellar remarks that the weakness of ancient science lay in its perception of analogies, and that Lucretius 'trusts too implicitly to their guidance.' As 'one prominent

nec satis est, proquam largos exaestuat aestus,
unde queat tantum suboriri ac subpeditare. ii. 1133-8.

[1] nequiquam, quoniam nec venae perpetiuntur
quod satis est, neque quantum opus est natura ministrat.
ii. 1141-2.

Here, as so often, Epicurus follows Democritus, who holds that the world increases so long as fresh matter from without is added to it, and wastes away so soon as this process stops. ἀκμάζειν δὲ κόσμον ἕως ἂν μηκέτι δύνηται ἔξωθέν τι προσλαμβάνειν (Hippol., 'Ref.,' i. 13).

[2] iure igitur pereunt, cum rarefacta fluendo
sunt et cum externis succumbunt omnia plagis,
quandoquidem grandi cibus aevo defit
nec tuditantia rem cessant extrinsecus ullam
corpora conficere et plagis infesta domare. ii. 1143-7.

[3] ii. 1150-74.

instance' of this influence, he points out that 'Lucretius's whole conception of the creation of the world is derived from a supposed likeness between the properties of our terrestrial and celestial systems, and those of living beings.' And again, 'in his explanation of our mundane system, he is both consciously and unconsciously guided by the analogy of the human body.' It appears to us that this comment on the passage which we have just examined is not altogether deserved. We must remember that the world, as Lucretius conceives it, is born like an animal, assimilates food and grows in bulk like an animal, and like an animal wastes away and dies. Thus it fulfils some of the chief conditions of an organism. Again, Lucretius conceives it as having pores in its outer circuit, the ether, 'as it were breathing-places all round it,' such as exist in the bodies of animals.[1] The earth (only a part of the world be it remembered, and therefore, according to Lucretius's notion of an organism, certainly not independently possessed of life)—the earth puts forth grass and bushes just as feathers and hair grow on animals,[2] while just as animals bear young, she has begotten from her surface all living creatures, and among them man, who was born from 'wombs,' developed on the face of the ground and attached to it by roots. Again, how strange is the union of the earth with the air and ether, all which three are 'conjoined and formed into a single being from

[1] undique quandoquidem per caulas aetheris omnis
et quasi per magni circum spiracula mundi
exitus introitusque elementis redditus extat. vi. 492-4.

[2] ut pluma atque pili primum saetaeque creantur
quadripedum membris et corpore pennipotentum,
sic nova tum tellus herbas virgultaque primum
sustulit. v. 788-91.

In the same sense (according to Karsten) Empedocles spoke of the earth's 'shaggy body'—λάσιον δέμας. Lucretius, moreover, follows Empedocles in the opinion that the earth produced plants first, and animals afterwards (Plut., 'De Plac. Phil.,' v. 26, 4).

the beginning of their existence,'—a union so perfect that the weighty earth lies without sinking in the bosom of the air, which holds it up in the same way as our soul, though so light, holds up the body.[1] (We must remember that Lucretius supposes the mind to be formed chiefly out of similar elements to the air and ether. Ancient thinkers very generally identified the substance of the human mind with the ether,[2] and though Lucretius denies that the ether is living or divine,[3] he certainly conceives it to perform, to some extent, the same part for the earth which the nervous system does for the human body.[4])

The world, as Lucretius conceives it, is at all events as much an organism as a tree,[5] perhaps not much less so than an animal. Since Lucretius supposes consciousness to accompany Organism when the latter is complex enough, we might seem to have some reason for inferring that he considers the world to be endued with some degree of consciousness so long as it lasts. How many of the ancient thinkers of Greece, the same men whose researches culminated in the atomic system, regarded the world as living and Divine! This conception was so widespread (partly, no doubt, because related to the old mythology) that few minds in antiquity were uninfluenced by it, as we see unmistakably illustrated in Lucretius's account of the origin of life and species, when in spite of himself he cannot help thinking of the earth as producing animals, exactly as if it were some great living creature. Lucretius

[1] This remarkable comparison is found at v. 556-63.

[2] We may compare the fragment of Pacuvius on the ether—

> Hoc vide, circum supraque quod complexu continet
> terram,
> id quod nostri coelum memorant, Graii perhibent aethera:
> quidquid est hoc, omnia animat, format, alit, auget, creat.

[3] v. 140-5, quoted below.

[4] See v. 554-63.

[5] Compare i. 774—

> Non animans, non exanimo cum corpore, ut arbos.

insists at length that 'earth, sun and heaven, sea, stars[1] and moon' are not divine, and are indeed 'so far from it that they ought rather to be thought a notable instance of what is destitute of vital motion and sense.' 'The mind cannot come into being,' he says, 'without the body, nor can it exist far from sinews and blood.' Thus Lucretius seems to anticipate Clifford. Because he can find no trace of a nervous system in any of the elements, he decides that it is not possible for Intelligence to dwell in any of them. Mind, he says, is found only in the living body, and therefore it cannot exist 'outside of the body and the living form altogether, in crumbling clods of earth, or in the sun's fire, or in water, or in

[1] It has been thought that Lucretius did conceive the stars as perhaps living creatures of a kind. His language is remarkable—

> sive ipsi serpere possunt
> quo cuiusque cibus vocat atque invitat euntis,
> flammea per coelum pascentis corpora passim. v. 523-5.

Referring to the Stoic belief that the stars are living, Munro says: '523-5 seems at first almost a Stoical doctrine; but is merely a poetical mode of saying that the fires of the stars are drawn on by that portion of the ether which provides them the fuel or nutriment they need,' and quotes Diog. L., x. 92. Most probably this is all the actual meaning, and Lucretius is merely indulging his fancy for personification. Still he speaks of the stars as if 'living' in two other passages—

> et tamen interutrasque ita sunt ut corpora viva
> versent v. 476-7.

> vivant labentes aetheris ignes. i. 1034.

Of course the passage v. 110-45 is conclusive against such a doctrine—

> tanto magis infitiandum
> totum posse extra corpus formamque animalem
> putribus in glebis terrarum aut solis in igni
> aut in aqua durare aut altis aetheris oris.
> haud igitur constant divino praedita sensu,
> quandoquidem nequeunt vitaliter esse animata. v. 140-5.

Yet according to Plutarch ('De Plac. Phil.,' v. 20) Epicurus held that the heavenly bodies are living.

the high borders of ether. Therefore these things are not possessed of divine sense.'[1]

But all these elements, we must remember, are in Lucretius's conception only parts of a whole—of one great Organism, the World, and he has before insisted that Life and Consciousness cannot exist in the parts of an Organism separately, but only in the whole.[2] Therefore he may without inconsistence[3] have conceived the world as, to a certain degree, a living creature. We merely point out here that Lucretius describes the world as an organism of a kind.[4]

2. Again, how does the world hold its place in the universe? This is a question which one cannot help asking. At first sight it would appear that Lucretius conceived the world to be eternally falling through infinite space,[5] and thus, though not

[1] v. 110-45. [2] ii. 904-14.

[3] While Plutarch, 'De Plac. Phil.,' v. 20, says that Epicurus believed the heavenly bodies to possess life, on the other hand, he asserts that Epicurus denied life to the world. 'Democritus, Epicurus, and all who believe in atoms and void, hold that the world is neither living nor is governed by Providence, but by "Nature" which is without intelligence (*φύσει δὲ τινι ἀλόγῳ*).'—'De Plac.,' ii. 3. Again, Stobaeus (p. 442, Gaisford) says, *οἵ μὲν ἄλλοι πάντες ἔμψυχον τὸν κόσμον καὶ προνοίᾳ διοικούμενον. Λεύκιππος δὲ καὶ Δημόκριτος καὶ 'Επίκουρος οὐδέτερα τούτων, φύσει δὲ ἀλόγῳ, ἐκ τῶν ἀτόμων συνεστῶτα*, this meaning (*τὸν κόσμον*), *ἐξ ἀτόμων συνεστῶτα, φύσει ἀλόγῳ* (*διοικεῖσθαι*). Both passages show how ancient philosophers regarded the atomic theory,—as essentially and of necessity atheistic. But whatever Lucretius's professed creed as an Epicurean convert may be, the world was something greater to him, in virtue of his gift of poet and of all therein implied, than it was or ever could have been to Epicurus.

[4] Epicurus's language might sometimes warrant a similar conclusion, as when he says that the world might perish after the fashion of a living creature, or of an animal. *Επίκουρος πλείοσι τρόποις τὸν κόσμον φθείρεσθαι· καὶ γὰρ ὡς ζῶον καὶ ὡς φυτὸν καὶ πολλαχῶς.*—Stob., 'Phys.,' p. 414. But perhaps these words merely mean that Epicurus, as his manner so often was, left this subject too an open question.

[5] See i. 1052-82, the text of which is unfortunately very imperfect. Munro's note is as follows: 'It is the Stoics, doubtless, whom Lucretius here mainly attacks, though the Peripatetics and some others held a

absolutely at rest, to be relatively so. Lucretius will not hear of any such principle as gravitation towards a centre of the universe. 'Through centre or no-centre alike heavy bodies must move downwards.' They cannot rest upon void, and it is against their nature to move upwards.[1] This explanation is the one approved by Lange.[2] Though it is the easiest and certainly the most apparent, still it is not the one most consistent with the doctrines of the Epicurean system. In the first place, it must be remembered[3] that Lucretius conceives the world to be exceedingly buoyant, since its heavy core, the earth, is surrounded by an outer envelope of ether and air, which, by their extreme lightness, diminish the action of gravity upon it, and 'cause the weight of the earth gradually

similar doctrine; they taught that there was but one finite world surrounded by an infinite void; and that the world was upheld in the way which Lucretius so clearly explains here, by things pressing to the centre: the earth resting ἰσοκρατῶς, in the words of Zeno in Stob., 'Ecl.,' i. 19, 4, at about the centre of the κόσμος, in the same way that the whole finite κόσμος remains fast in the infinite void. Had Epicurus, while retaining his conceptions of infinite space and matter and innumerable worlds and systems, seen fit to adopt this Stoical doctrine of things tending to a centre, and so to make his atoms rush from all sides of space alike towards a centre, he might have anticipated the doctrine of universal gravity.' Again, Mr. Munro remarks on ii. 251-93, 'In a curious memoir of the Berlin Transactions for 1782, by G. L. Le Sage, called "Lucrèce Neutonien," the author ingeniously argues that if Epicurus had had but a part of the geometrical knowledge of say his contemporary Euclid, and conceptions of cosmography the same as those of many then living, he might have discovered the laws of universal gravity, and not only the laws, but, what was the despair of Newton, its mechanical cause. Had he supposed the earth to be spherical, and made his atoms move in directions perpendicular to the surface of a sphere, that is towards its centre, he might not only have proved the law of the inverse-square of the distance, but have demonstrated the cause of that law.'

[1] Cf. the argument at ii. 184-215, immediately preceding the proposition as to the action of gravity. Lucretius here refutes the Aristotelian conception of a centrifugal upward motion.

[2] 'History of Materialism,' vol. i., c. 3.

[3] Chapter IV., § 4.

to pass away and be lessened.'[1] In the second place, as we have seen, Lucretius conceives the world to be kept in being 'by blows,' *plagis*—that is to say, by the constant supply of atoms rising up from below, and striking upon it from without. But plagae, the blows of the atoms, effect something more than this. Lucretius has made no positive statement on the point; but from one or two passages it may be inferred[2] that these blows of the atoms, constantly rising up from beneath and striking upon the earth, have the effect of keeping it suspended in its place in the universe. When we remember that Lucretius conceives the world as a whole to be exceedingly buoyant, and the natural gravity of the earth in this way to be greatly reduced, it seems clear that he supposed the upward impulse, given by the atoms ever ascending to feed the world, to be enough entirely to counteract the action of gravity upon it, and thus to keep it in its place in the universe, floating as it were in one spot amid the infinite sea of matter.

[1] Terraque ut in media mundi regione quiescat,
evanescere paulatim et decrescere pondus
convenit. v. 534-6.

[2] In the passage above referred to—

illud in his rebus longe fuge credere, Memmi,
in medium summae, quod dicunt, omnia niti,
atque ideo mundi naturam stare sine ullis
ictibus externis, i. 1052-5.

when we remember that Lucretius has just proved the necessity of matter rising from beneath, and that he is here refuting the Stoic doctrine that the world is kept in the centre of the universe by the pressure of all things towards that centre, it is difficult to resist the inference above expressed. It is evident that Lucretius has left this section of the poem unfinished, if not incomplete.

CHAPTER IX.

THE DE RERUM NATURA AT ONCE A POEM AND A WORK OF SCIENCE.

UNLIKE other books of which the subject-matter is mainly scientific, Lucretius's poem has a twofold aspect, one being its relation to science, and the other its relation to literature.

It is impossible to study the 'De Rerum Natura' without feeling that, entirely apart from its theological motive, Lucretius shows a keen intellectual passion for scientific research. It is a vivid delight to him 'to make his way into nature's secret hiding-places,'

caecasque latebras
insinuare omnis,[1]

and he glories in his master Epicurus having been the first to 'burst open the fast bars of nature's doors,'

effringere ut arta
naturae primus portarum claustra cupiret.[2]

Beyond question he shows much of the genuine scientific method in the manner according to which he pursues his inquiries. The modern scientific man may indeed smile at the air of rigid and unswerving scientific method which Lucretius sometimes assumes, while in the act of giving for some fact of nature an explanation which we can now easily see to be arbitrary and baseless. At the same time, Epicurean science deserves the credit of a careful and methodical examination of

[1] i. 407. [2] i. 70-1.

many natural phenomena, and of conclusions regarding them, which are as accurate as without the help of experiment could be looked for, and are often surprisingly near the truth.[1] If we take, for example, Lucretius's explanation of the cause of thunder, or of earthquakes, or of the eruptions of Aetna, we find a most praiseworthy accuracy in observing and collecting all the facts of such phenomena, and stating these simply and as they bear on each other. He next discusses the possible causes at work, explaining the mode of operation of each from what we daily observe in the world, and fixes shrewdly enough on the actual one.

These qualities are specially noticeable in his explanation of the way in which the magnet attracts iron.[2] He first reminds us that minute particles are constantly streaming off the surface of every substance. These emanations never cease. The efflux of particles from the magnet is quite unlike that from any other body. There proceeds constantly out of it an exceedingly violent current of atoms—so strong that it pushes out the air and creates a vacuum in front of the magnet. Then, when a ring of iron is placed near the stone, a great number of particles of the iron rush forward into the vacuum.[3] Since no other substance is formed of atoms more intricately entangled or more solid than those forming iron,—in consequence of this peculiarity of its atomic structure, when a large stream of its particles have moved forward, the whole mass cannot help speedily following. But why should the iron rush into the empty space? The reason is that the air 'beats on' (verberat)

[1] This does not hold good of Epicurus's notions about astronomy. Here he was far behind his age. Epicurus simply ignores the discoveries of Eudoxus, who was born a whole generation before him.

[2] vi. 906-1089.

[3] extemplo primordia ferri
in vacuum prolapsa cadunt coniuncta

Mr. Munro translates, 'forthwith the first-beginnings of iron fall headlong forward into the void in one mass.'

every substance on all sides, and wherever a vacuum is formed, the pressure on every side naturally forces the iron in the one direction where a passage is open. Moreover, the very resistance offered by the solidity of the iron makes the air more active in forcing it on. The air penetrates into its passages, and striking on its substance impels it forward from within as well as from without. 'The air makes its way subtly through the frequent pores of the iron even to its minute parts, and thrusts and pushes it on as does the wind a ship and its sails.'[1] In another way, too, the air inclines the iron to move. All substances are porous, and therefore must have air within them. This inner air is constantly moving to and fro, and makes a stirring in the substance of the iron. Thus, as soon as the iron has begun to move under the influence of the magnet, the motion of the inner air tends to carry it onward. Why, then, is iron attracted, but not other things? Some substances, like wood, are so porous that the streams of air, instead of pushing them forward, flow right through them. Lead, again, is too heavy to be carried forward in such a way.

Lange finds the following difficulty in Lucretius's explanation: 'How is it possible that the currents from the magnet can expel the air without repelling the iron by the same force?' But Lucretius has specially guarded against this very objection (at ll. 979-97). He points out that the pores of different bodies differ widely. 'The nature of the passages' allows one thing to pass, but refuses to admit another to enter the same substance. Thus the current from the magnet 'swims through'[2] the

[1] We may compare iv. 877-906, where Lucretius, in explaining how the human body is set in motion by the will, again assigns a secondary part to the air in assisting and increasing motion after it has begun. When the human body has begun to move, its pores are opened (how or why Lucretius does not say), and the air, which is ever ready to enter, flows in great quantity into the inmost parts of the body, and striking on its substance impels it onward, just as the wind beating on the sail drives on a ship.

[2] tranet (vi. 1052). So of 'images', which can pass through glass, because its pores are straight and parallel (Recta foramina tranant. iv. 601).

pores of the iron because there is a special adaptation between the structure of these pores and the particles of the magnetic current.[1]

After Lucretius has given us his explanation, no doubt we are still disposed to ask many questions. At the same time, as Lange has said, Lucretius deserves credit for explaining the magnet's operation not by the assumption of any mystical *horror vacui*, nor yet of any secret force or special sympathy, but as a consequence of the specific gravity and structure of the iron. This praise is well deserved. According to his fifth proposition, as we have seen, there is nothing but matter and void in existence. Therefore Lucretius is not misled by any notion of magnetism as an 'element' by itself, or a 'special fluid' or 'principle' of one kind or another. He treats the phenomena of magnetism *simply as properties of matter* residing in the molecules of the magnet. Here we see how great is the vantage-ground of Epicurean science.[2]

[1] According to Diogenes L. (ix. 47), Democritus wrote a treatise on the magnet which no doubt influenced Epicurus. Democritus, says Zeller, 'thought that the magnet and the iron consist of atoms of similar nature, but which are less closely packed together in the magnet. As, on the one hand, like draws to like, and on the other all moves in the void, the emanations of the magnet penetrate the iron and press out a part of its atoms which, on their side, strain toward the magnet and penetrate its empty interspaces. The iron itself follows this movement, while the magnet does not move towards the iron because the iron has fewer spaces for receiving its effluences' ('Pre-Socratic Philosophy,' vol. ii., p. 230, Eng. tr.). Democritus followed in the footsteps of Empedocles, who asserted the atoms of both substances to be similar, and 'supposed that after the emanations of the magnet have penetrated into the pores of the iron, and the air which choked them had been expelled, powerful emanations from the iron pass into the symmetrical pores of the magnet, which draw the iron itself and hold it fast' (Zeller, ib., p. 134). On this subject see Alex. Aphr., 'Quaest. Nat.,' ii. 23.

[2] Lucretius's explanation shows how his favourite doctrines, such as the streaming of emanations from all bodies and the porousness of all matter, could be made practically serviceable as working hypotheses. 'He dwells on the magnet at what appears so disproportionate a

One very characteristic feature of Lucretius's poem, in point of literary treatment, is the admirable power of illustration by which he can bring home to us a scientific doctrine through the means of something going on before our eyes in the world. Putting aside their beauty of poetic expression, these illustrations are in each case singularly close to the point. Their accuracy shows that, as Professor Tyndall has said, Lucretius possessed a strong 'scientific imagination.'

For example, this is how he illustrates the doctrine that the atoms are eternally in motion, even though matter appears to our eyes to be at rest: 'Thus often the woolly flocks as they crop the glad pastures on a hill, go creeping on whither the grass gemmed with fresh dewdrops invites and attracts each one, and the lambs full-fed are gambolling and butting each other in play;—all which objects appear to us from a distance to be blended in one, and to be like a white spot resting on the green hill.' A second illustration of the same doctrine is taken from the pageantry of war. 'Again, when mighty legions wage a mimic war and cover the plains with their swift movements, there the glitter rises to the sky, and the whole earth around gleams with brass, and beneath a noise is raised by the mighty trampling of men, and the mountains, smitten by the shouting, fling back their voices to the stars of heaven, while the horsemen hurry to and fro on the flanks, and suddenly charge across the plains, shaking them with their mighty onset.'[1] How vividly Lucretius makes us see the legions in evolution, with their armour flashing in the sun,

length, because the phenomena seem to him to illustrate so many of his favourite first principles. The elaborate criticism in Galen., 'De nat. facult.,' i. 14, of Epicurus's theory of the magnet, extending over many pages, proves that the latter must have dwelt on the subject at as great length as Lucretius does, and that he explained the phenomena in a similar manner' (Munro on vi. 917).

[1] ii. 317-30.

and hear their mighty tread! But still more impressive are the two following lines—

> et tamen est quidam locus altis montibus unde
> stare videntur et in campis consistere fulgor.

'And yet there is some place on the high mountains, from which they appear to stand still and to be a bright spot at rest on the plain.' In two short lines Lucretius shows us how small is all human activity, in mere bulk, as compared with the vastness of nature;—the mighty legions with their thousands of men constantly marching and counter-marching at full speed, seen a few miles away, appear only a small bright spot. Yet he could have chosen no better illustration to bring home to us how the tiny atoms may be in the most violent and constant motion, while the matter which they build up appears to us to be at rest.

While these illustrations are thickly scattered over the poem, when we come to examine them singly, we find that every one strictly illustrates the doctrine under which it is introduced. Lucretius enjoys weaving them in as pieces of pleasant colour into the somewhat grim framework of his poem —but not one is introduced merely as an episode. In each case the illustration makes plain to us a point which without it we should feel difficult to grasp. If we examine any one, we find how strictly it bears on the special doctrine under discussion. For example, that of the cow whose calf has been taken away to be sacrificed, and which keeps searching everywhere, returning again and again to wood and pasture and stall in the hope of recovering it. How does Lucretius introduce this, one of the most pathetic passages in all literature? It occurs in the strictest sequence. Lucretius is proving that the atoms are not all of one pattern, but that they are of many different shapes, and that this diversity explains why each individual of every species of living things differs from the other in appearance and form. Thus the animals recognize each other's individuality as much as human beings do—every parent knows its own offspring, and the cow robbed of its calf can never be satisfied by the substitution of any other,[1] but seeks persistently 'something special and known.'

[1] We quote Mr. Munro's admirable rendering of this passage: 'And in no other way could child recognize mother or mother child; and this we see that they all can do, and that they are just as well known to one another as human beings are. Thus often in front of the beauteous shrines of the gods a calf falls sacrificed beside the incense-burning altars, and spirts from its breast a warm stream of blood; but the bereaved mother as she ranges over the green lawns knows the footprints stamped on the ground by the cloven hoofs, scanning with her eyes every spot to see if she can anywhere behold her lost youngling; then she fills with her moanings the leafy wood each time she desists from her search, and again and again goes back to the stall pierced to the heart by the loss of her calf: nor can the soft willows and grass quickened with dew and yon rivers gliding level with their banks comfort her mind and put away the care that has entered into her, nor can other forms of calves throughout the glad pastures divert her mind and ease it of its care: so persistently does she seek something special and known' (ii. 349-66).

There must at bottom have been a kind and loving nature in the man who could read so keenly the signs of suffering in the animal creation as Lucretius has here done. Is there not something specially tender in the touch which tells us how not even the pasture-ground with its dear and familiar landscape—the drooping willows, the cool and dewy grass, and the river gliding past brimful—could make the cow forget her sorrow? If animals have any distinct love for the spot where they have been bred, surely it is bound up with just such associations of shelter, ease, and plenty. Too often we ignore the unhappiness which at times the lower creatures must suffer. Does not Lucretius actually dignify the suffering of the poor dumb creature, in its cruel bereavement, simply through the human sympathy with it, which he is not ashamed to express? In Lucretius's readiness to realize by imagination and to feel for the suffering, the pain, and the fear which must exist alike among men and among animals, all over a world like this where 'the whole creation travails together in pain,' there is an almost modern note. Thus, for instance, when he speaks of the lonely places of the earth, which are 'filled with quaking and terror' because of the fierce beasts of prey which the other creatures dread—

ita ad satiatem terra ferarum
nunc etiam scatit et trepido terrore repleta est
per nemora ac montes magnos silvasque profundas.

Lucretius's poem shows that he must have been a close observer of nature. The manner in which he musters facts from all quarters to illustrate any given doctrine shows what a grip he could take of general principles. Beyond question, in this way he contributed something to the science of Epicureanism beyond what he found in the treatises of his master. For example, when he is explaining the different sounds of thunder as the noise produced by the clouds, when buffeted or burst by the wind within them, he gives, as an instance of sound produced in a similar way, the loud flapping of a canvas awning over a large theatre, when it is tossed up and down by the wind, or again, the harsher sound heard when the wind tears it. Again, he shows in the following way that there are forces in nature which we cannot see except in their working.[1] First of all, he describes a violent storm, and points out that the wind, though its particles are minute, acts in exactly the same way as a stream in flood. Further, he says, we perceive smells, though we cannot discern their cause. Clothes hung up by the sea-shore become moist and are dried again by the sun, though we see no particles either come or go. In course of years the ring worn on a finger grows thinner, the dripping of water from the eaves hollows a stone, the pavement in the street is worn down by the feet of the passers-by. The hand of the brazen statue at the city gate is wasted away by the touch of those who salute it. The rocks by the sea-shore are eaten away by the salt spray, but we can perceive this gradual waste no more than we can discern the slow gradual increase occurring in the growth of plants.—In all these cases we are certain that things have become worn away, but 'the nature of vision' prevents us from seeing the particles which, at any given moment, disappear. Thus Lucretius proves that 'Nature works by unseen bodies,'

corporibus caecis igitur natura gerit res.

[1] i. 265-328.

These facts from Lucretius's daily observation by field and shore and street appear homely enough when thus abstracted, but they one and all bear rigidly on the point which he is illustrating. At the same time, how comes it that Lucretius's river in flood and torrent of storm in the air never remind us of the illustrations which are so plentiful in the modern science text-books? They are not mere diagrams to illustrate his text, but pictures drawn for us to see.

'The force of the wind when roused beats on the harbours and sinks huge ships and scatters the clouds; sometimes it rushes over the plains in a swift whirlwind and strews them with great trees and lashes the mountain-tops with forest-crashing blasts: so madly does the wind rave with shrill howling and rage with a threatening roar.[1] Winds therefore, beyond question, are unseen bodies. . . . They stream on and spread destruction abroad in just the same way as does the soft and yielding water when, in time of flood, down it rushes all at once in a torrent, while a great downfall of water from the high hills feeds it with copious rains, hurling together the fragments of the forests and whole trees: nor are the strong bridges able to sustain the sudden assault of the flood coming on,—in such wise does the river, wild with much rain, dash against the piers with mighty force, and overthrow them with loud crash, and roll along the great stones beneath its billows; wherever anything opposes its waves, it hurls it down. In this way, then, must the blasts of wind, too, rush along.'[2] Lucretius's picture of a river in flood is a perfect

[1] Or, if the MS. reading be correct,

saevitque minaci murmure pontus

'while the sea rages with a threatening roar.' The deep roar of the sea and the shrill howling of the wind are the two contrasting notes of the storm.

[2] i. 271-95. It is worth while to compare the powerful description of a storm from the hand of an English writer whose method of painting

example of a force at work according to natural law; but it contains also something more, something which we may find in Wordsworth, but for which we might for ever search the pages of the science-primers in vain. There is in it a touch of all that stirs the imagination in the grandeur of the flooded river. This wild torrent might pour down the hills of any region in Fairy-land, which the poets, who alone are the guides for that country, have explored for us. How few men have ever combined the scientific eye of the naturalist with a genuine sense of the surpassing beauty of the world's pictorial side! No gift is rarer. Many, indeed, presume to it, and many have attempted to unite both faculties; but in almost every case do we not feel this to be a mere presumption—something really ungenuine, however little meant to be so? If we put aside Goethe, possibly the only writer who has united a scientific instinct for the study of nature with a real sense of the beauty of the world in colour and form, is one recently gone from us, by no means gifted with Lucretius's intensity of poetic expression, but holding, in gift of qualities of

nature has, we think, much akin to that of Lucretius. It is from Charlotte Bronte.

'The wind . . . all day had blown strong and full from the south, without, however, bringing a speck of rain. Instead of subsiding as night drew on, it seemed to augment its rush and deepen its roar; the trees blew steadfastly one way, never writhing round and scarcely tossing back their boughs once in an hour; so continuous was the strain bending their branchy heads northwards—the clouds drifted from pole to pole, fast following, mass on mass; no glimpse of blue sky had been visible that July day.—It was not without a certain wild pleasure I ran before the wind delivering my trouble of mind to the measureless air-torrent thundering through space' ('Jane Eyre,' chap. xxv.).

This picture is not altogether unworthy of Lucretius himself. Lucretius, too, would have delighted in just such a night of storm as Charlotte Bronte here describes. Both feel the same kind of joy—like children watching round the hearth—in observing Nature, the great Housekeeper, at her work, whether clearing up in storm or at her everyday duties. Both of them are, in the same fashion, at home with Nature.

mind and character rarely united in the same man, a place in its way unique in literature—our own much-honoured Charles Kingsley. Almost any one of his books, from 'Yeast' to his 'Christmas in the West Indies,' will show what we mean. In neither Kingsley nor in Lucretius has the scientific standpoint driven out the sense of nature's beauty as something beyond all analysis. Thus in Lucretius's pictures of nature, however introduced, there is a touch which raises them beyond mere illustrations of scientific doctrine. Indeed, it is wonderful how easily, by a mere word or two,—like some great actor who can suddenly stir us to the heart by a simple change of tone,—Lucretius can at any time rouse our feeling of awe in the very midst of a scientific discussion. Thus, in the middle of a laborious explanation of the cause of thunder, the line—

Dant etiam sonitum patuli super aequora mundi,

at once brings before us the clouds moving slowly with solemn thunder high above the earth.

Lucretius's mode of painting nature is in some ways unique. He possesses the keenest eye for the pictorial, or rather the picturesque aspect of the world—that intense perception of the contrasts of light and shadow which is given only to the Latin race. We see it most notably in French literature, from the mighty hand which paints great and passionate pictures of tempest by sea and land in 'Les Travailleurs de la Mer' down to the simplest tale of story-tellers like Erckmann-Chatrian. The latter, if they describe but the interior of the meanest peasant's cottage, the coming-on of morning or of dark in a plain, unromantic village, or the early setting-in of winter among the Vosges, can make the scene stand before us vividly as if we saw the very spot, looked through the cottage window in the dark, or even felt the warmth of the logs crackling amid the deep silence of the frost. Lucretius, too, has this power, which makes us see the same landscape which he sees, and almost hear its sounds, and breathe its air with a vividness of picturing and lifelike projection which no other poet of the ancient world possesses.

Take again his picture of the gathering tempest, when the darkness vast as of hell fills the great caverns of the sky, while phantom figures hover overhead amid the gloom before the breaking of the storm—

> quod tum per totum concrescunt aera nubes,
> undique uti tenebras omnis Acherunta reamur
> liquisse et magnas coeli complesse cavernas:
> usque adeo taetra nimborum nocte coorta
> impendent atrae formidinis ora superne:
> cum commoliri tempestas fulmina coeptat.[1] vi. 250-5.

'For then the clouds close in a mass over the whole sky on every side, so that we might fancy that all its darkness had left Acheron,[2] and had filled up the great caverns of heaven: in such crowds do faces of black horror gather together amid the frightful night of storm-clouds, and hang over us from on high, at the time when the tempest is beginning to forge its thunderbolts.'

There is a touch here which no other poet can give us, unless perhaps it be Victor Hugo. Both he and Lucretius have at command words of the same strange and almost magic potency. Moreover, each of them constantly realizes the utter weakness of man amid the dread powers of Nature. And do not both poets convey to us the same sense of a background of tempest and terror which surrounds our human life?

We have spoken of the beauty of Lucretius's illustrations. It is necessary to note that every such apologue which de-

[1] The solemn motion of these lines shows consummate mastery of verse. The unusual succession of spondees recurring at ll. 250, 252, and 255 is certainly intentional. In particular, the concluding line at once calls up to the ear the long slow rolling of thunder just before a severe storm.

[2] *i.e.* the place of eternal darkness where the dead are. Acheron was a word of terror, like Cocytus and Styx, the very sound of which, says Plato ('Republic,' p. 387), 'makes all that hear them to shudder with fear.'

lights us by its beauty and freshness, is also a demonstration of some doctrine. 'There is properly speaking no episode in the poem,' says Martha. Lucretius's purpose is too keen and fixed to allow him ever to wander in any by-path, however tempting. Gladly would the reader follow him at times off his iron-bound pathway of Epicurean teaching to some detail of personal history, or of individual preference, or of the part played by himself in the busy world of Rome. But this he never does. Though a man of his keen and passionate nature must have had an eventful life, though the peculiar tone of his preaching against ambition and the mad Roman lust for power and pleasure might tell us that he too had played high for the prizes of the world in his day,[1] still he never allows himself to dwell on any details of his own past history. He is too absorbed in his mission of preaching salvation and peace, after the manner of Epicurus, to a suffering world. Not improbably, indeed, his conversion to Epicureanism did not occur till the later years of his life, and not long before he commenced his poem.

The English reader who has heard much of Lucretius's imagination and poetic charm is somewhat astonished when he finds the earlier part of the poem composed in great part of passages containing scientific argument and proposition of the most close and exact kind. When he comes to a passage like that beginning,—

[1] May we infer from iii. 170-4 that Lucretius had at some time been wounded in battle? At all events he analyzes the sensations following after a severe wound as if from personal experience.

> si minus offendit vitam vis horrida teli
> ossibus ac nervis disclusis intus adacta,
> at tamen insequitur languor terraeque petitus
> segnis, et in terra mentis qui gignitur aestus,
> interdumque quasi exurgendi incerta voluntas.

Again, in Book iv., he certainly speaks as one who had experience of passionate love, seemingly for one who was no very worthy object.

> quod si forte aliquis credit graviora potesse,
> corpora, quo citius rectum per inane feruntur,
> incidere ex supero levioribus, atque ita plagas
> gignere, &c.

and so on for many verses, in which Lucretius tries to prove that heavy bodies do not fall more quickly than lighter in the void, he naturally asks, 'Is this long scientific discourse poetry?' To this we would answer that the poem is penetrated through and through in its most severe and protracted reasonings, its plainest and most matter-of-fact statements, by the earnest purpose of the poet. It is this that turns the prose of it to poetry, and informs the plainest line with feeling. He frequently reminds us that the aim of his inquiry is not scientific, but to overthrow superstition. The atomic philosophy which he has stated with such care, and defined in its every doctrine with utmost exactness, is the foundation on which he hopes to build a system that shall deliver man from care and fear, and make it possible for him to live his life aright. This burning earnestness of purpose extends from the passionate poetry of the prologues even into the most arid discussions, so that even here Lucretius still holds the reader, and 'he cannot choose but hear.' But it is not merely earnestness—there is also a strong personal attraction in Lucretius. His chivalrous boldness draws us to him. The reader who studies him heartily follows by a kind of charm his figure in the front keeping his daring road, and does not desert the poet even when he leads through waste and stony places.

It is remarkable how completely Intellect and Imagination work in harness in the 'De Rerum Natura.' Nothing but the very strongest fire of feeling could have enabled him so to master such weighty subject-matter. No other poet has ever soared upward carrying with him so heavy a load, yet Lucretius does lift the burden which would chain other men to earth, and even make his relation of Epicurean doctrines in great part lucid and beautiful. Doubtless this strong white heat of his soul which alone could fuse such materials, made the writing of his poem in the last years of his life a most exhausting labour. He speaks of spending the quiet nights in composing it.[1] Even in sleep, he tells us, his mind still labours at the same task of investigating the nature of things, and striving to commit it to verse.[2] All through the poem he speaks like a man ever driven on by some consuming passion until the labour he has set himself is finished, and who is unable to take rest in any halting-place by the way. So strong and impetuous is the tide which rushes on from the first verse of the 'De Rerum Natura' to the last, that the energies of the poet may well have been wasted in its mighty and unfailing flow.

[1] quemvis sufferre laborem
suadet et inducit noctes vigilare serenas,
quaerentem dictis quibus et quo carmine demum
clara tuae possim praepandere lumina menti,
res quibus occultas penitus convisere possis. i. 142.

[2] in somnis . . . .
nos agere hoc autem et naturam quaerere rerum
semper et inventam patriis exponere chartis. iv. 965-70.

Compare ii. 729—

dicta meo dulci quaesita labore.

and iv. 419—

conquisita diu dulcique reperta labore
carmina.

CHAPTER X.

WHAT THE WORLD OWES TO LUCRETIUS. HIS TEACHING AND PERSONALITY.

THERE is much apparent reason to justify the belief that Lucretius's system was in reality atheistic. According to his master Epicurus the Gods exist, but in nature and function they are utterly unlike the Gods hitherto adored by mankind. They are to be worshipped because of their excellence, not because they have power to help or to hurt human beings. They have nothing to do in the world or for it, but live supremely happy and supremely idle in a stormless, cloudless Epicurean heaven which is situated in the *intermundia*,

> The lucid interspace of world and world,
> Where never creeps a cloud or moves a wind,
> Nor ever falls the least white star of snow,
> Nor ever lowest roll of thunder moans,
> Nor sound of human sorrow mounts to mar
> Their sacred everlasting calm.[1]

For Epicurus's atomic explanation of the world they are entirely superfluous. Why then, it is often asked, did not Epicurus, instead of thus pensioning off the Gods in the *intermundia* so as to be well out of the way, entirely give up these meaningless but pernicious shadow-deities? Is it not most probable that Epicurus really had no faith in any Divine existence whatever? The open profession of atheism would have rendered his system too unpopular (would have been 'dangerous,' as Lange says), and therefore is it not most likely

[1] From Tennyson's poem, 'Lucretius.'

that his profession of belief in the Gods was simply nominal and for appearance' sake, merely, as Lange calls it,[1] 'a concession to the existing order of things'? At present we can only reply that Lange's opinion is entirely without foundation. It is in opposition to that of scholars who, like Zeller, Munro, and Martha, have studied Epicureanism most thoroughly. There is unquestionable evidence that both Epicurus and Lucretius did believe, and believe firmly, in the existence of these Deities, strange as they are. What the relation was between the Epicurean and his new Pantheon is a subject which we cannot here enter upon. M. Martha[2] calls the religion of Epicurus 'a kind of obscure mysticism.'[3]

It is undeniable, however, that Epicurus left the Gods no part

[1] 'Geschichte des Materialismus.' Erster Abschnitt. Cap. iv. 1873.

[2] 'Une sorte de mysticisme épais' ('Le Poème de Lucrèce,' p. 106).

[3] We cannot help thinking that Lucretius, had he lived, would have concluded his poem with a description of the Epicurean Gods and their heaven. The poem, as it stands, stops abruptly in the course of a description of the plague. Lange says (p. 120) that Lucretius 'perhaps intentionally' concludes his work with a description of the power of death, as he begins with an invocation of the Goddess of Life. But we have absolutely no reason to suppose that Lucretius intended this for the conclusion. It seems plain (from vi. 92 ff.) that he intended the Sixth Book for the last. He has also expressly promised to describe the Gods and their seats 'at length' (largo sermone. See v. 146-55). As this promise is nowhere fulfilled, he must apparently have reserved this subject for the conclusion of the Sixth Book and of the poem. How then could Lucretius introduce such a subject after his description of the plague? Probably as follows:—The plague was supposed to come from the Gods and to be a Divine visitation—but it has been shown to come from natural causes. How should the Gods trouble themselves to produce such a thing, they who live in perfect calm and bliss? Such a connection of ideas is in accordance with Lucretius's way of introducing a fresh subject, and would harmonize well with the introduction of the book (ll. 50-79), proving that such interference is incompatible with the perfect bliss of the Gods. The picture of the Gods and their abode might possibly have occupied two or three hundred lines. Doubtless a wonderful piece of painting that would have been (for Lucretius has gorgeous colours at command), and not entirely uninfluenced by the old mythology.

whatever to play. And, when Lucretius says that the bodies of the Gods would be dissolved if the supply of matter were not infinite, is it not plain that their existence is less real to him than that of Matter? It is far more God to him than were his absurd, idle Epicurean Deities, who we need hardly say (though some living writers[1] approve and admire Epicurus's relation to the Gods) could not be Gods to him or to any man except in name. In reality, where Matter is conceived eternal, where it is able to evolve Life, where there is no Divine Being (or only a nominal one), Matter is there the supreme and ultimate reality. The mighty torrent of atoms, streaming for ever through space and capable of striking out worlds full of beauty and life by their combinations—this, in truth, is Lucretius's God.

It is thought by many that Lucretius's whole conception of the world is essentially irreconcilable with that of Divine action in any form. This is not difficult to understand when we think of such passages as the following—

> quas ob res ubi viderimus nil posse creari
> de nihilo, tum quod sequimur iam rectius inde
> perspiciemus, et unde queat res quaeque creari,
> et quo quaeque modo fiant opera sine divom.[2]

'Wherefore, when we have seen that nought can be created out of nothing, hence we shall now know more thoroughly that which we are seeking, namely, both the elements out of which everything can be produced and in what manner all things are done without the hand of the Gods.'

Or again:—

[1] 'On one great point the mind of Epicurus was at peace. He neither sought nor expected, here or hereafter, any personal profit from his relation to the gods. And it is assuredly a fact that loftiness and serenity of thought may be promoted by conceptions which involve no idea of profit of this kind' (Address delivered before the British Association, by John Tyndall, 1874, p. 10).

[2] i. 156-9.

Quae bene cognita si teneas, natura videtur,
libera continuo, dominis privata superbis,
ipsa sua per se sponte omnia dis agere expers.[1]

'If you well understand these things and keep them in mind, nature, free at once and rid of her haughty lords, is seen to do all things of herself and entirely of her own accord, without the aid of the Gods.'

Is not teaching like this distinctly anti-Theistic? it may be asked. Before this question can be answered, it is necessary to consider what notion of a Divine government of the world as well as of human affairs prevailed in Lucretius's day. Was it a noble or an ignoble one? Or does he write out of mere antipathy to any belief in Divine Providence, when he boasts that Nature works of herself 'without the hand of the Gods'?

If religion, in any genuine form, hardly existed in Lucretius's time, at all events the agency of the Gods in the world was most distinctly and constantly acknowledged. The whole fabric of Roman society and of the state was penetrated by a belief in what is called Divination, that is to say, that the result of every undertaking could be forecast, ere it was entered on, by inspecting the entrails of a victim, or by other kinds of augury,—moreover, that every event of importance, either to the individual or to the nation, was announced by prodigies of various kinds, and that to a certain extent the performance of proper sacrifices could avert any calamity thus foretold. It was not merely by the ignorant that such a belief was held, not by fishermen or shepherds only, but by the wisest and greatest in the land,—by the general as well as by the soldier in the ranks. The practice was officially recognized by the state. Any prodigy, as soon as it was announced, was examined into by the Senate, and officially reported on. Moreover, a college of augurs existed, whose business it was, on

[1] ii. 1090-92.

every important occasion, to ascertain the will of the Gods from various signs. 'Who does not know that this city was founded by auspices? that by auspices all things are conducted, during war and peace, at home and abroad?'[1] Auspices were taken at the election of a king, consul, or praetor, as well as of some other magistrates, and unless these could be reported favourable, the election was void. Without auspices no public assembly could be held. A general could not cross the frontier, or engage an enemy, unless the birds sanctioned. Crassus's defeat by the Parthians was explained by his having fought against the auspices. Pompey was compelled to fight at Philippi because the auspices were favourable.[2] Frequently too, private affairs, marriage for instance, were not entered into without thus trying to elicit some sign of Divine approval. Besides the flight of birds and the examination of victims, there were, as Pliny tells us, eleven different kinds of lightning, each with a different signification. Such notions regarding natural phenomena, we need hardly say, rendered any scientific view of nature utterly impossible. If the harvest failed, it was not from improper farming so much as because some one of the divinities concerned had not been properly propitiated. When hail destroyed the olive-orchards of the Athenians, it was because some god wished to gratify their enemies, the Spartans. 'According to ancient religion,' says Martha, 'there is no law, everything in nature is arbitrary and disconnected, phenomena depend on Divine caprice, the thunderbolt, eclipses, movements of the heavenly bodies, the simplest things, the flight of a bird, the running stream, adeo minimis etiam rebus prava religio inserit deos. There is no physical science, there is only one art of value, that of augurs

[1] 'Auspiciis hanc urbem conditam esse, auspiciis bello ac pace, domi militiaeque omnia geri, quis est qui ignoret?'—From the speech of Appius Claudius, Livy, vi. 41. Compare Cicero in Vat. 6, 'auspicia, quibus haec urbs condita est, quibus omnis respublica atque imperium continetur.'

[2] Both instances are cited by M. Martha, 'Le Poème de Lucrèce,' p. 362.

and diviners, since at any instant all may be confused by the bad-humour, the benevolence, or even the forgetfulness of the divinity concerned. . . . One may say that in the eyes of Pagan credulity, not only do no physical laws exist, but there are no political and moral laws. When Rome and Carthage are at war, and their fleets are going to encounter, the people in suspense ask not on what side is right and justice, on what side the generals are the best and the soldiers the most valiant, in a word, on what side virtue is, but whether the sacred chickens will consent to eat.'[1] We can easily imagine what the moral consequences of such a system must have been,—on one side, the lying in the name of God and premeditated trickery practised by a priestly class, living by the basest fraud, and, on the other side, the childish fears and the utter undermining of Conscience caused by a system according to which the Divine approval was secured, not by the sole qualification of righteousness, but by minute attendance to forms of propitiation.

And what was the character of these deities so busy in the world? Probably the divine Beings pictured in Virgil's Aeneid are not unlike the Gods of the popular belief. How then does Virgil represent them? False, treacherous, revengeful; they never forget or forgive, but watch with feline vigilance for an opportunity to punish those who have all innocently offended them. To the average Roman, not learned or speculative, the world was full of invisible powers, multiplex, jealous, ever-watching. The more zealously that a man paid attention to one God, the more might he offend another, since each different circumstance of life had its own guarding Deity. It was difficult to ascertain the will of each; it was easy to offend without knowing it. Thus Lucretius paints the anxiety of the devotee to leave no altar unhonoured

omnis accedere ad aras.

Again, if anyone were careless and paid no heed to the Divine

[1] M. Martha, 'Le Poème de Lucrèce,' pp. 73-4.

intimation, however trifling, what terrible unknown consequences might follow. Even a dream, however ludicrous, may possibly be a warning of some coming evil. The springs of religious observance, as Lucretius knew it, were indeed little else than Fear.[1] No wonder that he says that 'human life lay crushed under the heavy burden of superstition,' that men, like children in the dark, dread things which do not exist at all, and that he rejoices in the thought that 'nature has been rid of her haughty masters.' Doubtless there were many at Rome who professed utter disbelief in any Divine Being, but these were even more superstitious than the rest. Such men had the most complete faith in all kinds of portents and in astrology.[2] According to Divination every occurrence in Nature bore some reference to man—it came not from a natural cause, but from some immediate supernatural impulse, which again resulted from the Divine approval or disapproval of some preceding human action. What a confusion was this! How could man ever walk in safety and peace of mind amid a world so ordered?

[1]
quippe ita formido mortalis continet omnis,
quod multa in terris fieri coeloque tuentur
quorum operum causas nulla ratione videre
possunt ac fieri divino numine rentur. i. 151-4.

cetera quae fieri in terris coeloque tuentur
mortales, pavidis cum pendent mentibu' saepe,
et faciunt animos humilis formidine divom
depressosque premunt ad terram propterea quod
ignorantia causarum conferre deorum
cogit ad imperium res et concedere regnum. vi. 50-55.

[2] We do not here refer to the detestable Oriental worships which about this time began to be introduced at Rome, to which Lucretius himself bears witness,—

unde etiam nunc est mortalibus insitus horror
qui delubra Deum nova toto suscitat orbi, v. 1165-6.

nor yet, as with spiritualism in the present day, to the increase of magic and necromancy among those who had cast off religion. Most true is the word of Novalis, 'Where Gods are not, spectres rule.'

Surely the Romans owed a debt to Lucretius for establishing the fact of natural law, and thus breaking down the basis on which such a system was built. And some did not refuse to own this gratitude to the poet, if we may judge from the exclamation of Virgil, who does not add Lucretius's name—probably only because he dared not—

Felix qui potuit rerum cognoscere causas.

'Blessed is the man who was able to learn the causes of things!'

We are thus compelled to conclude that in Lucretius's day the popular and prevailing notion of Divine action was an utterly debasing as well as a false one. We need hardly say that a religion like this, however rigidly it may have insisted on formalities of worship, was utterly without true reverence. 'The impious man,' said Epicurus, 'is not he who denies the existence of Gods like those commonly worshipped; on the contrary, the impious man is he who asserts the Gods to be such as the vulgar conceive them.'[1] This famous sentence we see to be no mere controversial figure of speech, but literally true. Such a conception of the Gods had to be broken down and utterly destroyed before it was possible to have any worthy notion of a Divine Power. It is not wonderful that Lucretius becomes boastful in relating the victory won by his master and himself over religion, that he waxes over-confident in the strength of his own arm. He feels that he has escaped from superstition as from imprisonment in a gloomy stifling house with low-roofed chambers and contracted windows. He has cut his way through the phalanx of his priestly gaolers, and now is in the open air. Who shall say that there he is not nearer God than in the foul prison-chambers of Roman superstition? His poem is the voice of one calling to others still fast bound to break forth and come into the light. Lucretius might use the very words of Goethe:—

[1] ἀσεβὴς γὰρ οὐχ' ὁ τούτους τῶν πολλῶν θεοὺς ἀναιρῶν, ἀλλ' ὁ τὰς τῶν πολλῶν δόξας θεοῖς προσάπτων.—Diog. L., x. 123.

Away, away from the darkened rooms,
  Where they grudge you the light of day,
Where men low-bowing in craven fear
  To their misshapen idols pray.
Of superstitious worshippers
  Enough in the years of old! To-day,
Have done with portent, myth, and ghost.
  Leave them all to your teachers grey.[1]

Lucretius is indeed over-confident[2] in the power which has

[1] Gesetz der Trübe, from Goethe's 'Gott und Welt.'

[2] But is Lucretius ever so positive as one or two modern writers—Professor Clifford, for instance? There is nothing in Lucretius even approaching the concentrated bitterness with which Clifford regards every kind and degree of religious belief. 'If we once admit that physical causes are not continuous, but that there is some break, then we leave the way open for the doctrine of a destiny or a Providence outside of us, over-ruling human efforts and guiding history to a foregone conclusion. . . . I do think that, if it is right to call any doctrine immoral, it is right so to call this doctrine, when we remember how often it has paralyzed the efforts of those who were climbing honestly up the hill-side towards the light and the right, and how often it has nerved the sacrilegious arm of the fanatic or the adventurer who was conspiring against society' ('Lectures and Essays,' vol. ii., pp. 59, 60). Time is hardly given to man for the purpose of breaking on the wheel such a passage as this, but did Lucretius ever write anything so recklessly partial in statement? This modern Lucretius has indeed compassion for those against whose belief he is compelled to utter the awful verdict of science, yet we cannot help hearing a finer note of pity, more tender, more human, which sounds from long ago,—

O miseras hominum mentes, O pectora coeca!
qualibus in tenebris vitae, quantisque periclis
degitur hoc aevi quodcumquest! . . . .
nam veluti pueri trepidant atque omnia coecis
in tenebris metuunt, sic nos——

'Only for another half-century,' Clifford in one passage imagines his hearers to pray,—"Only for another half-century let us keep our hells and heavens and gods."—It is a piteous plea'—piteous indeed, but the laws of nature are not more inexorable than Professor Clifford. No! he says, you cannot keep any longer 'these sickly dreams of hysterical women and half-starved men.' He concludes with a solemn warning, 'Take heed lest you have given soil and shelter to the seed of that awful plague [*i. e.* Christianity] which has destroyed two civilizations,' and so on (vol. i., delivered him, namely, the science of Epicurus. His combat with superstition has been too recent and too fierce for him to recover calmness. He is still panting from the struggle with 'Religion who showed her face from heaven, frowning upon mortals from on high with awful aspect.' By the side of this dread form, grim and cruel, he might have drawn another picture which might embody the very spirit of his poem—the fearless figure of Science, sword in hand, exulting in her triumph, and with a light not altogether born of Heaven upon her face.

It is well to remember that the world owes to Lucretius the first firm and clear exposition of the doctrine of Law and steadfast universal order in Nature. This principle is, we need hardly repeat, a direct inference from his atomic philosophy; if all the productions of nature obey fixed laws, the ultimate elements of the matter from which they are composed must be definite and unchangeable.

> materies quia rebus reddita certast
> gignundis e qua constat quid possit oriri.[1]

The whole scientific teaching of his poem is a sermon in expansion of the text—

> et quid quaeque queant per foedera naturai,
> quid porro nequeant, sancitum quandoquidem extat.[2]

'It is absolutely decreed, according to the conditions of nature, what each thing can do, and what it cannot do.' 'Nature is conquered by obeying her.' True indeed is the old maxim, but how can we obey her, unless we first learn her laws, and how can these be known unless they are permanent and stead-

p. 253). At sight of any phenomenon we inquire for its cause,—Ex nihilo nihil. Where the philosophical student finds a Lucretius, he will look for an Iphigenia near by. What are the errors of doctrine or of practice which have inspired the words of this modern Lucretius, who is so much more bitter than the ancient one? Mr. Pollock mentions in his preface (p. 31), a fact as to Clifford's religious opinions while at Oxford which may, to a certain extent, explain his attitude as a reaction from an opposite extreme.

[1] i. 203-4.

[2] i. 586-7.

fast? The very knowledge that there is a uniform order in Nature gives man a feeling that Nature is less formidable, that he has gained a certain power over her and is less at her mercy, and that as he acquires more knowledge of her ways, he will continually secure for himself greater safety and well-being. Without unswerving laws of nature, how could the world be a training-place for man at all? without them what could we learn?—indeed we should be ever unlearning. Lucretius's poem taught this great lesson to the European races, and on this account his name ought always to be spoken with gratitude.[1] It is his consciousness of the greatness of this truth which gives dignity to the tone of his poem, even when it presents to us old and outworn explanations of the phenomena of nature, and where otherwise its subject-matter is wearisome enough. For the ignorant multitude life, he tells us, is a struggle in the dark, but 'the clear sight of nature and her law' frees man from all childish terrors where no cause of fear is, and teaches him to find his safety in obedience to the world's great Order,—

> hunc igitur terrorem animi tenebrasque necessest
> nec radii solis neque lucida tela diei
> discutiant sed *naturae species ratioque.*

Though Lucretius asserts the realm of Law and preaches that the welfare of man lies in obeying it, still he has no weak confidence in Nature. Indeed he shrewdly suspects that Nature, on the whole, is not friendly to man. Storm, earthquake, and pestilence, show that Nature does not care and will lift no finger, turn no step aside from her steady course, to save from destruction her proudest offspring, who is 'the roof and crown of things.' The relation between Nature and man is not friendship, but may rather be called an armed warfare. Epicurus, he says, 'showed what evils existed everywhere in

[1] Mr. A. Benn, in his able work on Greek philosophy, has attempted to deprive Lucretius of this honour. See Appendix.

the affairs of men . . . because nature had so ordained, and from what gates you must sally out to encounter each.'[1] Thus Epicurus delivered man by teaching him not only nature's laws to be obeyed, but also the rules of scientific warfare and the proper defence against each attack of nature's besieging forces. Nature, he says again, seems to trample on the pride of man as if in scorn of human weakness. 'When the utmost fury of the violent wind sweeps along the sea the admiral of the fleet, with his mighty legions and elephants, does he not approach the Gods with vows to seek their mercy, and terror-stricken ask in prayer for a lull in the winds, and favouring gales?—in vain—since none the less many a time is he seized by the furious hurricane and borne on to the shoals of death; *so utterly does some hidden power trample on the greatness of man*, and seems to tread under foot the renowned fasces and cruel axes, and to make sport for itself of them.'[2]

Some write as if the mere recognition of Law in nature, along with the determination to obey it, must of itself bring joy and peace to the mind of man, even though he utterly disbelieve in any Divine Mind whose will that law expresses. On this principle Lucretius's clear perception of a regular Order

[1] quidve mali foret in rebus mortalibu' passim,
quod fieret naturali varieque volaret
seu casu seu vi, quod sic natura parasset,
et quibus e portis occurri cuique deceret. vi. 29-32.

[2] Munro supposes that Lucretius may here be thinking of the consul M. Claudius Marcellus, lost at sea just before the third Punic war.

summa etiam cum vis violenti per mare venti
induperatorem classis super aequora verrit
cum validis pariter legionibus atque elephantis,
nec divom pacem votis adit ac prece quaesit
ventorum pavidus paces animasque secundas,
nequiquam, quoniam violento turbine saepe
correptus nilo fertur minus ad vada leti?
usque adeo res humanas vis abdita quaedam
opterit et pulchros fascis saevasque secures
proculcare ac ludibrio sibi habere videtur. v. 1226-35.

everywhere pervading the world must have brought him peace and gladness. But is it so? Too true it is that the regularity of natural law is not always admirable. Nature, we see, will take the utmost, nay, did we judge of her as we should of a human antagonist, the most cowardly advantage of our weakness or ignorance. Nature 'has no pity; for some awful but most good reason, she is not allowed to have any pity.'[1] Those who blindly admire the unswervingness of nature without realizing that, even when cruellest, she must work to some far-off end,—*to those whose creed allows them to see no further*, this regularity must in one important sense appear base and not noble. If a great machine, swinging its polished rods in accurate time, strike the life out of a beautiful child which ventures near it, so far it is a stupid machine. It has come into collision with a far higher organism than itself and destroyed it, blindly and to effect no higher end, but simply because it cannot help doing it. In so far its regularity is stupid and insufficient. No man who has a heart for human suffering can logically admire the regularity of natural law in all its working, unless he admits that (according to the simple and beautiful allegory of one who has passed from us), 'Madam How' (that is, Nature and the law) is in all her ways the servant of 'Lady Why.' But Lucretius did not believe in the existence of any 'Lady Why,' or if he did believe, could not persuade himself to own it. He has explained Nature as a great machine working with a perfect accuracy which he admires, but he is too honest to deceive himself about a world where he sees suffering on every side, and sees man punished by penalties beyond measure for the sins of his fathers. One does not worship a machine because it is vastly strong and works regularly. No! there is something far finer in its essence in the brave little spark within the heart of the man who plunges into the whirling waters to save life from a wreck, than there is in all the appalling vast-

[1] Charles Kingsley.

ness of the storm, when sea and tempest, working with a mechanical regularity even in their utmost rage, hurl thousands of tons of water crashing on the fated ship. Lucretius is keenly conscious of this. He has far deeper admiration for the heroic in man, for the fearless spirit that will dare death, or calmly face supernatural terrors for the sake of his fellows, than he has even for the great *Natura rerum*. This is seen in his fervent worship of Epicurus, 'who ventured first to lift up his mortal eyes to the face of religion and first to withstand her to her face:' 'A God he was, a God, most noble Memmius, who first found out that plan of life which is now called Wisdom,[1] and who by his skill rescued life from so great billows and from so deep a darkness, and stationed it in so perfect a calm and in so bright a light.'[2] The man who could feel such emotion and such admiration for another could never worship a machine, however perfect.

Experience enough Lucretius may have had (from disease, as some think) that Nature is pitiless. Certainly his unfinished poem (cut short when just in sight of the goal,[3] but without reaching it) points to an activity suddenly broken off—by what we cannot tell, not improbably in consequence of some severe nervous disease, to which his hints of frightful visions haunting his waking hours as well as his dreams, seem to point. Thus it is possible,—whatever it be that may have gone wrong in his life,—that Lucretius may himself have been caught between the grinding wheels of his great machine, the machina mundi, which, regular to a fault, could not and did not spare

[1] Lucretius thus appropriates the name *Sapientia* for the teaching of Epicurus. Before Epicurus no Wisdom was. For this passage we may compare the words of Plutarch (quoted by Munro), Ἐπίκουρος τἀγαθὸν ἐν τῷ βαθυτάτῳ τῆς ἡσυχίας ὥσπερ ἐν ἀκλύστῳ λιμένι καὶ κωφῷ τιθέμενος.

[2] i. 66-7; v. 8-12.

[3] tu mihi supremae praescripta ad candida calcis
currenti spatium praemonstra, callida musa. vi. 92-3.

him. In any case, he shows himself to us as no blind worshipper of Nature—

> nequaquam nobis divinitus esse creatam
> naturam rerum: tanta stat praedita culpa.[1]

'The Gods can never have created nature: so vast are its faults.' So to our own Shelley the sounds of Nature are at times the voice of the world lamenting over its pain:—

> Sad storm, whose tears are vain,
> Bare woods, whose branches strain,
> Deep caves and dreary main
> Wail for the world's wrong.

A keen-sighted human spirit cannot have its eyes closed with darkened glasses of any kind, not even with scientific and brand-new Epicurean spectacles, which assure him that all things are atomic, but that all is well and the best that can be. A soul like Lucretius's cannot be hoodwinked thus,—even though it may sternly compel itself to assent on logical conviction. Beyond question Lucretius's creed did much to darken his life. There is a want of hopefulness in his poem. Nor is this wonderful. Is it not true that our pleasures are in great part made up of anticipated repetition? The pleasure of to-day is sweet, and we trust that we shall enjoy the same to-morrow and to-morrow in extended series. Nay, we even feel that repeated enjoyment shall be keener,—

> To-morrow shall be like to-day, but much more sweet.[2]

Is not this anticipation perhaps the finest part of pleasure? So, if to all life there be no great To-morrow, is not a shadow cast over all that we most keenly enjoy, whether work or recreation, in our daily life? These things are once, but shall be no more. There is nothing to anticipate regarding them. Is it possible that a man's mental horizon, all around, should not be narrowed and darkened by this? It is the human heart by which we

[1] v. 198-9; repeated at ii. 180-1.

[2] From Christina Rossetti's 'At Home,' a little poem, in its own way, worthy of Dante ('Goblin-Market and Other Poems,' 1865).

live, and Lucretius has denied two of its deepest instincts. So all that is most cheering and glad in human life and human hopes is, for him, maimed and curtailed. He seems to wonder why other men find life so sweet. Frequently, in reading his poem, we feel that we are in presence of a nature somehow isolated and shut off by itself, so that it cannot rightly share in common human cares and human pleasures.—In the poet's own past life there may have been special reasons for this, besides the very plain ones just indicated.

So far as a determined assent to logical conviction is implied, few men have ever more completely accepted the gospel of Atoms and Eternal Death than did Lucretius. Whatever moments of truer consciousness he may have had, he will own no higher power ordering the world than the blind eternal motion of the atoms. He has sought and found nothing but 'dead matter' anywhere. And, as it has been said of some recent philosophical theisms, that they place God 'somewhere in the nebula of Orion,' entirely out of connection with earth, so Lucretius's shadow deities are put far, far away from him in the *intermundia*—

> The shadowy interspace of world and world—

and thus the earth and heaven were left empty. Not so was it with the old mythology in its better day. His master Epicurus had driven away the beautiful and glorious Beings, the Heroes and Gods, who, as the simpler and purer forefathers of his race believed, would stoop from heaven and walk with man upon earth, caring for his cares, watching his efforts, ever guiding and helping him. He had banished the many conscious living Presences from sky and field and wood and stream. He had found the world living, if not with One Great Life, yet with many lower lives, and he left it dead, 'utterly dead' and cold. This was his substitute for polytheism, and the pagan races, as a whole, could never have accepted that substitute, for it was not more but less true. And do we not find that the poem of his disciple Lucretius strikes us forcibly with a sense of *something wanting?* of a missing element alike in the world of man and in the outer world of nature? Yes, even in those vivid descriptions of nature, which he knows so well to paint in a line or two of his rushing verse, pictures so bright and clear and living that the fresh air seems to blow out of them as on a breezy summer morning, even here we feel that something is absent, something which we know ought to be there, and at first we are puzzled and ask ourselves what it is. It is just so in a dream, when we go out and walk the familiar street and see the well-known forms and persons, the trees, the sky, the house in which we live; but always we are filled with pain, for however bright the sunlight, and however green the leaves may be, there is something wanting in the landscape—everywhere—in all we see. It is some friend, someone that we dearly love, and without whom we can never be satisfied—whom in our dream we go seeking up and down yet cannot find, and go about in misery because of it. And what at times strikes us so strongly in Lucretius's poem, is something more than his mere air of standing aloof from common human affairs. Is it not the lack of Life, the absence of any Living Presence in the world that we feel? And alas! this was no dream to Lucretius, but a waking nightmare—and to those who can hear, the sadness of it cries aloud in his poem.

Lucretius is a most ardent disciple of his master. None more zealous or more determinedly true to his great teacher could well be. The system of Epicurus held him with a mighty grasp, and he schooled himself to crush down all instincts that rebelled against it; but sometimes human nature was strong within him, and Epicureanism was weak. When he stood in presence of the beautiful world, Lucretius felt that there was something there which was wanting in his own darling 'De Rerum Natura.' From many a passage in his poem we cannot help believing that in presence of Nature and her life, Lucretius was stirred by feelings directly at variance with the dogmas of his sect. Then he seems to have felt happier; then he forgot himself and his Epicurean creed, forgot the bitter combat with religion, forgot that human life, for all save a few, 'is a struggle in the dark,'[1] even forgot altogether to anticipate what he had compelled himself to think of as merely a blessing and end of troubles,—

> The first dark day of nothingness,
> The last of danger and distress.

At such a time the dogmas of Epicurus slide away, vanish and are gone. Again the world is glad to him as in his childhood. In such a mood we imagine him in some spot which he loved to haunt, perhaps in a meadow on the side of the sunny Apennine, on a day in spring. The sun is beating down on the green grass, and there is a scent of trees budding, and in the pasture lambs and calves are sporting or running to their dams, 'intoxicated in their young hearts by the pure new milk,'[2] joyful creatures, without fear or suffering, all happy and each living a life of its own[3] in the lap of mother Earth. In every blossoming hedge young birds have their nests, in every pond and rivulet living things are swimming to and fro, to all of whom existence is happiness. There is joy all over the world. And in the midst lies Lucretius, with the branches swaying over his head,

> prostratus . . . In gramine molli
> propter aquae rivum sub ramis arboris altae.

To him also Nature is kind. She has prepared form and colour and scent and sound, and mingled them to satisfy his every sense. And he, too, is glad for the season; why—he can hardly tell, but he feels that behind all this beauty and

[1] omnis cum in tenebris praesertim vita laboret. ii. 54.

[2] hinc nova proles
artubus infirmis teneras lasciva per herbas
ludit, lacte mero mentes perculsa novellas. i. 259-61.

[3] See ii. 333-80.

manifold animal gladness there lies, like some boundless sea, a deeper life, vast and happy. It is thus that passages of his poem speak to us, telling of moods in which his Epicurean creed could barely, as if for duty's sake, throw a veil over the throbbing life of the world which bursts through the covering and will not be hid. Over and over again he has told us that the world is chance-born, but at the same time he says to us through grand and lovely pictures of nature, 'See! is not the world more mysterious, and does not its beauty compel our awe more than anything of mere atomic parentage could do?' Most true and apt, therefore, are the words of Mrs. Browning; Lucretius

Denied
Divinely the divine and died,
Chief poet on the Tiber-side.

Even while he most stoutly denies that there is any Divine element in the world, this poet of Materialism rouses us to feel that there is something in and behind its grandeur and its ever-changing beauty which rebukes our sensual impulses and compels our worship. In short, Lucretius, like the writers of the Psalms, makes us feel that there is a spiritual element in Nature.—To attribute such thoughts to Lucretius may be termed mere fancy,—simply without foundation, if not contrary to fact,—as those who know the poet merely as an ardent Epicurean may tell us. But Lucretius was not merely an Epicurean, he was also most genuinely a poet, with a poet's heart, that is to say, with a heart more human than that of other men, and therefore one which craved more deeply to assert its relation to God as well as to man, *and specially one to which the beauty of the world, of necessity, carried even stronger and more certain inspirations than it does to other men.* Such a nature must have broken into rebellion now and then, and listened to the voice of its deepest instincts, which at other times were rigidly chained down.

Lucretius has indeed most minutely explained to us the Birth of the World from chance atomic combination as a matter with no mystery about it. He has shown how the wheels of the machine were first set in motion, how it keeps going on, and how steadily it works. So far he follows his master. Yet, in spite of this, in his inmost thought and feeling it is clear that the world is no dead mechanism to him. He does not regard it, as the ordinary man of science would now have us do, as a mere eight-day clock, a marvellous time-keeper. Nor yet is the world to Lucretius, as the roué deems it, a battered old machine, grim and grey, coming wearily up to time for yet another spring. No! Lucretius cannot help thinking of it in the way that all great poets have done, as something divinely beautiful and grand, which, even though his own life may be sad, still remains to him fresh and bright as it is to the child.

It is remarkable that a system of atomic materialism should allow the existence of Free-will at all. Epicurus and Lucretius, however, most firmly believed that man has free choice and is responsible for his actions. It is this alone which dignifies Lucretius's system, based as it is on Materialism, and which enables us to respect him as a moral teacher. If there is no Free-will, there can be no such thing as sin. Each of us, sensual and cowardly, or pure and heroic, but plays the part to which he was fated by his birth. 'It cannot be denied,' says Martineau, 'that the whole system of moral conceptions, feelings, and language, rests upon the belief in Free-will, and deals with man as (within its particular range) the real cause of what he is and does. But for this, who could suffer compunction any more for a lie than for a squint, or shame for *delirium tremens* more than for a typhoid fever, or feel more indignant disgust at the crimes of a Caesar Borgia than at the rapacity of a wolf. Remorse for sin would be impossible but for the consciousne that it lies at our door.'[1] Lucretius's firm belief in man's freedom, in the fatis avolsa potestas, 'the power wrested from the fates,' gave him a strong conviction of human responsibility for rectitude or guilt. Thus, he tells us, it is in our power to eradicate our own evil temperaments,—not indeed completely, so that one man shall not still be more prone than another to anger or fear, but training and reason will do almost everything.[2] 'So very small,' he concludes, 'are the traces left of men's original natures, which reason is unable to expel from us, that nothing hinders us from living a life worthy of the gods.' Consistent with this belief is Lucretius's keen conviction of the misery brought by guilt to the sinner. Repeatedly he dwells on the power of conscience to torture the guilty soul, as, in his picture of the youth captivated by a courtesan, who feels that he is squandering his years and his wealth,

conscius ipse animus se forte remordet.[3]

He repeatedly points out that the criminal, even though his sin be not found out, cannot live a happy life. 'Violence and wrong catch all who commit them in a net, and for the most part recoil

[1] 'Ethics and Religion,' by James Martineau. 1881, p. 12.

[2] Lucretius attributes these differences of disposition to differences in the atomic structure of the mind—yet even so, he deems man the master, and not the slave, of his own inherited temperament.

quamvis doctrina politos
constituat pariter quosdam, tamen illa relinquit
naturae cuiusque animi vestigia prima . . . .
inque aliis rebus multis differre necessest
naturas hominum varias moresque sequacis;
quorum ego nunc nequeo caecas exponere causas
nec reperire figurarum tot nomina quot sunt
principiis, unde haec oritur variantia rerum.
illud in his rebus videor firmare potesse,
usque adeo naturarum vestigia linqui
parvola quae nequeat ratio depellere nobis,
ut nil impediat dignam dis degere vitam. iii. 307-22.

[3] iv. 1121-40.

on him from whom they sprung.'[1] We seldom find in Lucretius any direct teaching which shows a consuming passion for personal righteousness. On such points his expressed teaching rather takes a negative form; it aims, that is, at saving men from the fear and the evil appetites which make their lives misery. Yet there is a fervour of what is, in reality, genuine religious feeling in Lucretius. Thus, in denying the common notions of hell and its torments, he points out that in truth hell is not outside a man, but within him. Trûly it consists in doing wrong and the consequent misery of mind. 'There is in this life a dread of punishment for evil deeds, signal as the deeds are signal, and as an atonement for guilt there is the prison and the frightful hurling down from the rock, scourgings, executioners, the condemned dungeon, the boiling pitch, the red-hot plate, the torches. Yea, and though these be wanting, still the conscience-stricken mind (mens sibi conscia factis), fearing before the time, applies to itself goads, and frightens itself with whips, and sees not meanwhile what end there can be of ills, nor what limit at last there can be to punishments and fears, lest the same evils be increased after death. *The life of the fool*, in short, *becomes a hell here upon earth.*'[2]

Hic Acherusia fit stultorum denique vita.

[1] circumretit enim vis atque iniuria quemque
atque, unde exortast, ad eum plerumque revertit,
nec facilest placidam ac pacatam degere vitam
qui violat factis communia foedera pacis.
etsi fallit enim divom genus humanumque,
perpetuo tamen id fore clam diffidere debet;
quippe ubi se multi per somnia saepe loquentes
aut morbo delirantes protraxe ferantur
et celata mala in medium et peccata dedisse. v. 1152-60.

If we compare with these lines the passages quoted by Munro (on l. 157) from Epicurus, we find that Lucretius adopts a different tone from his master's in referring to the consciousness of guilt.

[2] sed metus in vita poenarum pro male factis
est insignibus insignis scelerisque luella,

There is far more that is spiritual in the denial of hell and its terrors from such a standpoint than there was in the orthodoxy of the poet's day. Lucretius has the utmost contempt for merely formal religion.. At thought of this, a sudden blaze of anger, mixed with pity, flashes out of him. 'There is no holiness in being often seen to turn oneself with veiled head towards a stone, and to approach every altar and to fall prostrate on the ground, and to spread out one's palms before the statues of the gods, and sprinkle the altars with much blood of beasts, and link vow on to vow,—rather is it to be able to look on all things with a mind at peace.'[1] Lucretius sees most clearly that religion consists not in making oneself safe, and in anxiously striving to propitiate the Gods. He finds more of true religion in an open-eyed soul which looks the facts of the world full in the face, without any slavish dread of unseen powers—

pacata posse omnia mente tueri.

The anxiety to appease the Gods by devout ceremonies springs

carcer et horribilis de saxo iactu' deorsum,
verbera, carnifices, robur, pix, lammina, taedae;
quae tamen etsi absunt, at mens sibi conscia factis,
praemetuens adhibet stimulos terretque flagellis
nec videt interea qui terminus esse malorum
possit nec quae sit poenarum denique finis
atque eadem metuit magis haec ne in morte gravescant.
hic Acherusia fit stultorum denique vita. iii. 1014-23.

[1] nec pietas ullast velatum saepe videri
vertier ad lapidem atque omnis accedere ad aras
nec procumbere humi prostratum et pandere palmas
ante deum delubra nec aras sanguine multo
spargere quadrupedum nec votis nectere vota,
sed mage pacata posse omnia mente tueri. v. 1198-1203.

Lucretius here refers to the most solemn forms of Roman worship; '*velatum* refers to the Roman custom of praying velato or operto capite' (with head covered), and *vertier ad lapidem* to another posture in worship, 'the suppliant approached in such a way as to have the statue of the God on his right, and then, after praying, wheeled to the right so as to front it, and then prostrated himself.'—MUNRO.

from Fear, and Lucretius seems to have felt strongly that selfish Fear has little in common with true worship. He does not attempt to explain this conviction, but is it not, even from his own standpoint, profoundly true? Especially, however, we cannot help seeing a religious tone in the intense earnestness with which Lucretius often warns men against the slavery to ambition. At such times he seems to single out his hearer, and to speak with a voice not out of centuries past, but as it were standing by our side. He laments that men 'wear themselves out to no end, and sweat with blood as they toil along the narrow road of ambition.' 'Avarice and the blind lust of honours compel hapless men to step across the bounds of right, and, sometimes as accomplices and agents in crimes, to strive night and day with surpassing toil to struggle up to the heights of power.'[1] From the safe and stable high places, 'well fortified by the learning of the wise,' to which Epicurus has raised him, Lucretius looks down with deepest pity 'to see other men wandering all abroad and going astray in their search for the path of life, to see the contest of intellect, the rivalry of birth, the striving night and day with surpassing effort to struggle up to the height of power and be masters of the world. O miserable minds of men! O blinded hearts! in what darkness of life and amid how great dangers is passed this little term of our existence, all that there is of it.'[2] There are many such passages. Why is it that such warnings affect us quite differently from those of Horace to much the same effect? Because it is a different nature that speaks to us, because there is far more of a spiritual tone, and not merely selfish prudence in Lucretius. He does not say, 'See the foolish toiling of these ambitious men, who might have done better, risked no battles by land or sea, spent no nights in watching, lived many safe

[1] v. 1131-2; iii. 59-63.

[2] ii. 7-16. We quote almost literally from Mr. Munro's vigorous rendering.

and easy days.' No! At thought of this, Lucretius feels a pity touched with admiration. It is the earnest effort and dauntless striving which he admires,—but wasted—which he pities, and cannot pity too much, seeing that men spend their money for that which is not bread. In Lucretius's preaching against ambition we discern a passionate and daring soul which finds it hard to curb itself; in Horace we see a timid and restricted nature which has not courage to attempt the heights. In no other utterances does Lucretius's personality become so distinct as in these. It may well be that his frequent warnings against ambition are the voice of one who has himself toiled 'along the narrow road,'

angustum per iter . . . ambitionis,

and who has found that such a life can in no case satisfy the hunger of the soul. Few have ever felt more deeply that the cause of man's unhappiness is within himself, that even with all earthly success, and with every luxury at command, he is still disquieted. And for this, Lucretius tells us, the heart of man is to blame. He needs some saviour from without to cleanse it, and to point out to him the viam vitae, 'the way of life.' By a striking parable he shows that we need to change ourselves, and not our circumstances. The heart of man, he says, is like a vessel which is corrupt and befouls all the precious things which are poured into it, or which is 'leaky and full of holes, so that it can never by any means be filled full. . . . Therefore,' he says, Epicurus 'cleansed men's hearts with his truth-speaking precepts and fixed a limit to lust and fear, and showed what is that chief good which we all are seeking, and pointed out the road along which, *by a short cross-track, straight and direct,* we might struggle towards it.'[1]

The brilliant prologues to Lucretius's books are simply throbbing with the most intense enthusiasm for humanity. In

[1] atque viam monstravit, *tramite parvo*
qua possemus ad id *recto* contendere *cursu*. vi. 27-8.

each of them Lucretius warns his readers with pathetic earnestness against ambition,[1] and points out the only way of salvation, that brought by Epicurus. Thrice he concludes with the same beautiful and tender words. It is reason alone which can deliver men from care and fear, 'since the whole of life is a struggle in the dark. Just as children in the dark tremble and dread every object, so we in broad daylight fear, sometimes, things which are no more objects of terror than those which children shudder at in the darkness and fancy they must certainly exist. This terror, therefore, and darkness of the mind must be dispersed, not by the rays of the sun and the bright shafts of day, but by the aspect of Nature and her laws.'[2] A small and selfish nature does not bear such pity for others of human kind. Again, his admiration for the spiritual bravery of Epicurus, and his supreme scorn for the cowardice of soul which, for fear of evils in this world or in the next, dare not look facts in the face,—is this not the sign of a fearless spirit and of loyalty to truth? Further, we have shown how keenly Lucretius makes us feel the grandeur and beauty of the world. Beyond question, it is never in the mirror of a selfish and sensual mind that the world can reflect itself thus nobly. Lucretius vaunts himself so boldly as the poet of Materialism that he has often been

[1] Especially in the magnificent prooemium of Book ii., ll. 1-61. Probably in all literature there is no other so penetrating description of the nature of ambition. Lucretius here sounds the deepest places of the human heart.

[2]
omnis cum in tenebris praesertim vita laboret.
nam veluti pueri trepidant atque omnia caecis
in tenebris metuunt, sic nos in luce timemus
interdum, nilo quae sunt metuenda magis quam
quae pueri in tenebris pavitant finguntque futura.
hunc igitur terrorem animi tenebrasque necessest
non radii solis neque lucida tela diei
discutiant, sed naturae species ratioque. ii. 54-61.

Repeated at iii. 87-93 and vi. 35-41.

thought to be a man without any religious instincts. But do we not find the certain signs of a spiritual nature in the features here indicated,—in his utter hatred of cowardice and falsehood, in his deep pity for the suffering of mankind, and in his passionate eagerness to give to other men the salvation which he has found for himself?

We are now better able to estimate the position of Lucretius and the aim of his poem. We see that he was a man of intensely earnest temper. The vision of Nature had filled his soul with the majesty of natural law. To him Nature seemed far grander than the old gods of the Pantheon at their mightiest. Moreover, he could not but feel that the conscience-nature of man with its abhorrence of wrong and cruelty represented something infinitely higher than the old impure, selfish, jealous Gods. Conscience, too (though he misunderstood its origin and the source of its authority), told him that they were false. It is difficult to decide whether Lucretius is to be viewed primarily as the opponent of Paganism or as a physical inquirer, whether his strongest craving was to pursue science or to cast out the superstitious terrors of a false and insufficient creed. It is indeed probable that, had he not been so deeply impressed with the evils of the national religion, and had he not seen in the atomic theory a philosophical weapon against them, he would never have drawn the subject of his poem from such a source. In his age there were many who found little difficulty in accepting this theory as a proof that the Gods have not created man, and, so far as he is concerned, are powerless for good or for evil. Lucretius, too, accepted the scientific system of Epicurus, and followed it up with all the strength of his intellect,—the more so as he had a natural faculty and decided fondness for such pursuits. It was a dogma of Epicurus that physics has a right to exist only for the sake of ethics, in order to show the falsehood of superstition, and that for any other end such inquiries are useless. If Lucretius, a man of far more eager and earnest temper, held this opinion at all, it was in a far less absolute form. In the study of nature for its own sake he found the keenest pleasure. Still it is true that even the strong intellectual passion which he shows for scientific research pales before the intense white heat of his human sympathies. Perhaps these are nowhere more strongly shown than in the wonderful description of the sacrifice of Iphigenia. Of all the Greek legends dealing with human sacrifice, this was by far the most painful and repulsive. No self-offered victim she, according to the legend, but one dragged to the altar, weeping and all unwilling. She is impelled by no high resolve to drive the knife into her own breast, to give her life of her own will for the sake of her countrymen. She simply feels that she is young and that life is sweet. There is no bright calm halo of self-sacrifice here that might move us to forget the horrible superstition that made cowards of the standers-by, warriors though they were, but fearing, far worse than death, some awful judgment from offended Gods, should they dare to snatch the knife from the hand of the priest and bid the victim go. Who that has once read can ever forget Lucretius's description of the weeping human victim—the young girl decked with the fillet on her soft hair,[1] like a beast for sacrifice, dropping on the ground in terror when she sees the approving priests, who stand by and conceal the knife, appealing in vain to her father, and at last carried by force to the altar? The scene is painfully vivid. Probably Lucretius may have seen horrible punishments inflicted at Rome for

[1] cui simul infula virgineos circumdata comptus
ex utraque pari malarum parte profusast. i. 87-8.

For us these lines have no force, but to a Roman reader they must have been most impressive. The victim made ready for the sacrifice, with woollen fillets hanging down at equal length on either side of its head, was a common sight in Lucretius's day. Iphigenia is the human victim, decked beast-wise. How would our own Shelley have done justice to a subject like this!

offences against religion.[1] At any rate, he uses this story of the past because he believes that the religion of his own day is fit to produce evil deeds and crimes like this, and does produce them.[2] If he had drawn but this one picture, its every detail

[1] If Lucretius had not himself seen the punishment by burial alive of a Vestal virgin, at all events he had heard it described by eye-witnesses. About the year 114 B.C., Aemilia, Licinia, and Marcia suffered thus (Dio Cassius, fragm. of Book 34). The offence so punished was unchastity, which might be real, or was presumed of the priestess under whose charge the holy fire had gone out. In such cases, the chief priest, the Pontifex Maximus, after praying silently with outstretched hands, led the victim to the entrance of the cell (cubiculum subterraneum.—Pliny, 'Ep.,' iv. 11). She then descended by a ladder, the cell was shut by the executioner, and earth piled closely over it—a horrible death! The execution was carried out in public, and, as Plutarch tells us, the populace watched it in awe-stricken silence. The place was 'near the Colline Gate, on the right hand of the road' (Livy, viii. 15). Lucretius must often have passed this hideous spot, justly called 'The Accursed Field,'—Campus Sceleratus.

[2]
quod contra saepius illa
religio peperit scelerosa atque impia facta. i. 82-3.

Lucretius is not thinking of distant times only, nor yet of far-off barbarous lands like Gaul and Carthage in his own day, where such rites were frequent and often many victims slain at once. Human sacrifices were not unknown even at Rome, and in Lucretius's own time. After the terrible disaster at Cannae, four victims (two Greeks and two Gauls, of either sex) were buried alive in the Forum Boarium, at a spot where such sacrifices were repeatedly performed;—'a most un-Roman rite,' as Livy calls it (minime Romano sacro, Livy, xxii. 57). Again, two years after the birth of Lucretius, with reference to some human sacrifice which had been carried out publicly, the Roman Senate passed a decree forbidding such rites (ne homo immolaretur, Pliny, 'Hist. Nat.,' xxx. 3). They still subsisted however. *Etiam nostra aetas vidit*, remarks Pliny, who died in A.D. 79 ('Hist. Nat.,' xxviii. 3). Plutarch (A.D. 46-120) refers to them in a tone of horror as still performed during the month of November (ὅις ἔτι καὶ νῦν ἐν τῷ Νοεμβρίῳ μηνὶ δρῶσιν Ἕλλησι καὶ Γαλάταις ἀπορρήτους καὶ ἀθεάτους ἱερουργίας.—Marcellus, iii.). A little later the Emperor Hadrian had to issue an edict against them. Even in mild and civilized Greece human sacrifices were offered, as, for instance, to Zeus, on the summit of Mount Lycaeos in Arcadia (referred to in the Platonic Minos, p. 315). According to Porphyrios ('De Abstin.,' ii. 23), human victims were offered on this spot even in the end of the third century A.D.

speaking his burning abhorrence of cruelty in religion's name, he had not lived in vain. Indeed this seems to us the noblest, bravest thing that he was allowed to do. Surely when man seeks to propitiate Deity and win his favour by sacrificing his weaker brothers, this is the incarnation of selfishness. Human self-seeking can go no farther. What could Lucretius do but protest against a power like this? The bare picture is enough, but his feeling rises to a climax in the single concluding word,—

Tantum religio potuit suadere malorum!

Could there be a God, and leave this appeal unanswered? Not in Lucretius's day did the answer come, not till years after he had died, perhaps, as tradition murmurs, by his own hand and in a moment of despair. Yet an answer did come, and the next generation saw it.

It was not for nothing that there was put into the heart of this man a burning hatred of the wrong done in the name of God, which Paganism authorized, and of the baseness of superstitious fear. In this picture of Iphigenia's death the horror that filled him at the thought of the evil deeds and deadly selfishness that superstition had prompted and produced, found as it were a mute expression. Short indeed, but trenchant like a thrust of the Roman sword, is Lucretius's comment on this shameful deed,—

Tantum religio potuit suadere malorum!

With this one blow he stabs the cowardly figure of cruel Superstition through and through. The echo of that terrible stroke is ringing through the world still.

Viewed in a wider horizon, and with reference to the progress of the world, we may look at his poem and even say, 'It is well and rightly done,' yet not altogether well for Lucretius himself, for he had done violence to the God-consciousness within him! His aim was to show that the ancient religion, which assigned for natural operations irregular, capricious Divine agents, was contradicted by the newly-discovered majesty and regularity of nature's laws, while the conscience of man remonstrated against the cruelty and wickedness which it sanctioned. And beyond question the poem must have had a mighty power in destroying the old polytheistic creed, which could never be made new again and had to pass away. Especially for the young who were thoughtful and earnest, who shared Lucretius's horror of the evils produced by superstition, who were moreover convinced by the intellectual force with which he refuted the notion that everything in nature goes by Divine caprice, and who, like him, could stifle neither of these convictions,—for such readers its passionate and impetuous poetry must have driven its arguments home with terrible force. Not in vain had Lucretius raised against Paganism a voice which could never more be silent.

We cannot help here referring to the estimate of Lucretius expressed some time ago[1] by a writer who has done some admirable work in the way of criticism, Mr. Robert Buchanan. 'Lucretius,' he says, 'is a materialist, pure and simple, solemn and staunch; as bigoted in his creed and as certain of his gospel as the veriest divine that ever thumped a cushion;' and so on. Nowhere has Mr. Buchanan hinted that Lucretius took a great step forward for religion as well as for science when he asserted that everything in nature goes by law and nothing by caprice; but passing over this entirely, he shows not the least appreciation of the fact that Lucretius was forced into his position by a recoil (an exaggerated recoil, of course) from the national religion of his day, with all its necessary results of sin and degradation. He talks knowingly of those 'modern writers who would fain make him (Lucretius) a mere enemy to the ancient polytheistic religion.' But does Mr. Buchanan know what this polytheistic religion was? What its moral in-

[1] 'Lucretius and modern Materialism.' 'New Quarterly,' April, 1876.

fluences were? Of what kind was the conception of Deity which Lucretius revolted from? In short, has he ever spoken a word against any conception of God which assigns to Him goodness and universal power in the world? Lucretius had not to do with the (in many respects) noble and beautiful worship of Greece at its best time. Paganism in his day had become a religion of sensuality and superstitious dread. As we have seen, the Gods of Rome, according to the popular conception of them, could inspire only repugnance and fear. In Mr. Buchanan's hands the poet becomes a 'mere' enemy of religion, impelled solely by a fanatic zeal against any and every spiritual conception. From this point of view, his earnestness in preaching Epicureanism simply shows that (as Carlyle has said of an eagerly proselytizing atheist), he 'was eaten up by the zeal of the devil's own house'—nothing further! And this means that he was a *mere* monster; for one who avows Lucretius's creed for simply these reasons is nothing more than this. This estimate of Lucretius is barely consistent with the facts of human character, and is moreover thoroughly unhistorical.

In truth Lucretius did good service to religion. It was commonly believed in his day that the world is a machine, with many hands tampering with it, sometimes stopping it, and then setting it going again. But surely such a notion is utterly fatal to any worthy conception of Divine action. The belief in a regular Order in Nature, in one Power at work instead of many, was like a great breath of cool air, bringing calm where before all was confusion and the alarm of utter uncertainty. It is necessary for men to realize that there is but one power at work around them, before they can form any notion of Divine action which will not be at once refuted by the manifest facts of Nature. It is true that after the old notion of many Divine agencies had been discarded, the conception of Nature as a new self-working Power might easily come, and Lucretius can hardly help at times speaking of Nature as if he denoted by it some active force or agency. Still we must beware of attributing to the poet ideas which are foreign to him.[1] Nature, as he intends to use the word, means only the laws of Nature, the habits of the world, that is all. Lucretius, indeed, had toilsomely levelled the road and prepared the way by which men might mount up to enjoy a truer conception of God in His relation to the world ; but though he made a path for others coming after him, he never ventured himself to ascend by it. He could not say, as does our own dear Chaucer—

Lo! I Nature,
Thus can I form and paint a creature,
When that I list; who can me counterfete?
Pygmalion? Nought though he alway forge and bete (prepare)
Or grave or paynte, for I dare well sayn
Apollo's Zeuxis shoulde wish in vain,
Either to grave, or paynte, or forge, or bete,
If they presumed me to counterfete.
For He that is the Former principal
Hath made me his Vicar General
To form and paint earthly creature
Right as me list: all things are in my cure
Under the moone that may wane and waxe,
And for my work nothing will I axe.
My Lord and I are fully at accord.[2]

When Lucretius wrote—

Natura videtur
ipsa sua per se sponte omnia . . . agere,[3]

he rose to a grand conception. Yet he would have been horror-stricken at the thought of taking the further step, and affirming that while the old Gods were false, Nature is but the Will

[1] For instance, Mr. Symonds says that Lucretius 'dropping the phraseology of atoms, void, motion, or chance, spoke at times of Nature as endowed with reason and a will,' but no one of the passages which he quotes (v. 186, 811, 846) appears to have this meaning.

[2] From the 'Tale of the Doctor of Physic.'

[3] ii. 1090-2.

of the one God at work. It was reserved for a great religious thinker of Latin race to say, 'Dei voluntas natura rerum est.'[1] Augustine was a man humble enough and wise enough to learn from those most violently opposed to himself in opinion, and we know not what light he may have owed to Lucretius's poem.

If Lucretius never shared Augustine's belief, we must remember that such a thought seems never even to have occurred to him. Possibly, had he lived longer, and had his mind grown free enough from the iron yoke of his master, he might have seen that the new conception, Natura, necessarily involved the existence of One Divine Power, omnipresent and all powerful. But as we see him in his poem, it appears as if he could not conceive of omnipotence in Deity. In one passage he expressly denies it—the universe is too vast, he says, to be governed by the Gods; no one can conceive of a Deity who 'can be present in all places at all times.'[2] Elsewhere he speaks as if it were equally absurd to attribute omniscience to the Gods. One cannot help contrasting with such passages the Hebrew conception of Jehovah, infinitely powerful and infinitely wise.

'Who hath measured the waters in the hollow of his hand, and meted out heaven with a span, and comprehended the dust

[1] Augustine, 'De Civitate Dei,' xxi. 8.

[2]
nam pro sancta deum tranquilla pectora pace
quae placidum degunt aevom vitamque serenam,
quis regere immensi summam, quis habere profundi
indu manu validas potis est moderanter habenas,
quis pariter coelos omnis convertere et omnis
ignibus aetheriis terras suffire feracis,
omnibus inve locis esse omni tempore praesto. ii. 1090-1104.
et simul in multas partis qui credere possis
mittere? an hoc ausis nunquam contendere factum,
ut fierent ictus uno sub tempore plures?
at saepest numero factum fierique necessest,
ut pluere in multis regionibus et cadere imbris,
fulmina sic uno fieri sub tempore multa. vi. 411-16.

Compare also the whole paragraph vi. 379-422.

of the earth in a measure, and weighed the mountains in scales, and the hills in a balance? Who hath directed the Spirit of the Lord, or being his counsellor hath taught him?

'With whom took he counsel and who instructed him? .... It is he that sitteth upon the circle of the earth, and the inhabitants thereof are as grasshoppers; that stretcheth out the heavens as a curtain, and spreadeth them out as a tent to dwell in .... To whom then will ye liken me, or shall I be equal? saith the Holy One. Lift up your eyes on high and behold who hath created these things, that bringeth out their host by number: he calleth them all by names by the greatness of his might, for that he is strong in power; not one faileth.'[1]

To the same effect writes many an old Psalmist,[2] who feels moreover that, much as he has thought of God's working, he still knows infinitely little of it. 'Behold this is a part of His ways, but how small a portion is known of Him!' Lucretius, on the other hand, cannot think of God in his relation to the world except as a man-like artificer. He never realizes Divine action except in the narrowest and feeblest way. Can Lucretius be excused in this? By no means. Distinctly and emphatically he is to be blamed in that he entirely cast away the best thought of the wisest who had gone before him. He simply ignores the teaching of Socrates, of Plato, and many another Greek thinker. He does not even argue against such views, which conceive the Divine nature as omnipotent and all-wise; he merely leaves them out of account. Might he not have learnt from Socrates, who, as Xenophon tells us, 'held that the Gods take care of men, though not in the fashion that the many believe. For these think that the Gods know some things, but others they do not know, but he

[1] Isaiah xl.

[2] Compare Ps. cxxxix. 6-9, xcv. 3-5, lxv. 5-8, cxxxv. 5-7, cxlvii. 4-5, and especially the whole of Ps. civ., where the Psalmist shows how the whole world has its being in God.

considered that the Gods know all things, both what is said and what is done and what is meditated in silence, and are present everywhere, and give intimations to men concerning all human affairs.'[1] Lucretius, indeed, deserves the rebuke administered by Socrates to Aristodemus, 'It becomes you, therefore, not to think that while your own eye can reach many furlongs, the eye of the Divinity is unable to see all things at once, nor yet that your mind is able to think about things here and about things in Egypt and in Sicily as well, but that the mind of the Deity is unable to take heed at the same time of all things.' In truth, he tells him, 'the Divinity is so mighty and of such a nature as to see all things and to hear all things at once, and to be present everywhere and take care of all things at the same time.'[2] The very school whom Lucretius so bitterly opposed, the Stoics, weak as their physics were, how much grander and more adequate than his was their conception of God,—of one universal Providence which governs all things. No doubt Cleanthes was an author despised by Epicureans; yet Lucretius must at some time have read his grand 'Hymn to Zeus.'[3]

'O Zeus, thou most glorious of the immortals, many-named, at all time almighty, O founder of nature, who guidest all things according to law. . . . Thee this whole world, revolving around the earth, obeys whithersoever thou guidest it, and is willingly governed by thee. . . . Without thee no

[1] Xenophon, 'Memorabilia,' i. 1, 19. [2] Ibid., i. 4, 17-18.

[3] Stobaeus, 'Ecl.,' i., p. 30. Cleanthes, famous even among the many remarkable figures of Greek philosophy for his laborious and earnest life, was originally a boxer, but came to succeed Zeno as chief of the Stoic school (about 260 B.C.). For years, while studying under Zeno, he chose to support himself by the severest bodily labour, as a water-drawer at night in the gardens of Athens. His contemporaries considered him by no means brilliant, and reproached him with lacking 'boldness' as a thinker. Heroic figures like his remind us of the intense moral earnestness—a passion veiled by a never-failing habit of self-command —which marked the Greek character.

work is done on earth nor in the sphere of the divine ether nor on the sea, except what the wicked perform in the foolishness of their minds.'

There is an enthusiasm in this hymn which Lucretius could hardly share, a glad feeling of loyalty, a delight in the law of God which is happiness to those who obey it, but misery to those who do not. 'The wicked, ever desiring to get gain, see not, neither do they hear the universal law of God, obeying which with their mind they might win a happy life.' Cleanthes' closing prayer to Zeus to scatter from the soul the 'dire ignorance' which makes men prefer ambition or greed or the pleasure of the body to righteousness, the entreaty 'to share in the understanding in the might of which thou, O father, dost with justice govern all things,'—such an aspiration shows a higher craving for personal goodness and purity than we anywhere find in Lucretius's poem, which (though full of nobleness and burning with the desire to help humanity) springs from a lower level. Lucretius has no such faith in one supreme Providence ordering all things—δίκης μετά—'with justice.' He has only the

> varios conexus, pondera, plagas,
> concursus, motus, per quae res quaeque geruntur,

'the various entanglements, weights, blows, clashings and motions of the atoms by means of which things severally go on.' This is Lucretius's substitute for Providence, while his creed is the *Nam certe neque consilio*[1]—'Verily not by design.' But just as we often see nowadays that the professor of cruel and gloomy doctrines is unselfish and even hopeful in character, happily Lucretius also is in many ways better than his creed.

[1] nam certe neque consilio primordia rerum
ordine se suo quaeque sagaci mente locarunt,
nec quos quaeque darent motus pepigere profecto,
sed quia multa modis multis primordia rerum,
ex infinito iam tempore, &c. v. 419-27.

The whole passage is translated at p. 56.

No reader of Lucretius's poem can help being struck by his frequent expressions of gratitude to Epicurus. We cannot doubt that these are genuine and from the heart, and that the teaching of Epicurus had brought to him help in his time of need. If, then, the poet regarded Epicurus as his deliverer, it may be asked,—From what had Epicurus saved him? When we read Lucretius's poem, we cannot doubt that he owed to Epicurus more than a merely intellectual debt,—more than the light of science,—more, too, than deliverance from superstitious fears. Lucretius speaks with the voice and the emotion of one who has known from personal experience in some form the miseries of a life of ambition,—who has known, too, the mad Roman lust for luxury and pleasure, and for ever new pleasures, and who has himself felt the 'sad satiety' that comes from drinking deep of such delights. He writes like one who (to use his own image) watches, safe on shore, the vessels in peril, in the grip of the tempest, and vividly remembers how he himself was once helplessly tossing on the same sea. His eagerness of temperament tells of a nature not easily curbed. The poems of Catullus show what a vortex of sensual life was the Roman capital in Lucretius's day.[1] Into that vortex it may be that Lucretius had sunk far deeper than ever Catullus did. For him it was impossible to float so lightly on the top of foul depths as did that bright and rare spirit. We can imagine Lucretius brought by his own eager and passionate nature into some dark crisis of disappointment and despair,—far more readily than one like Catullus, whose spirit was more buoyant and of swifter recovery. Is it impossible that at such a crisis Epicureanism, of the nobler kind, may have become known to

[1] Munro believes that Lucretius's poem was published scarcely a year before the death of Catullus, and that many passages in the 'Peleus and Thetis' show that Catullus had read it. See his note on iii. 57; also Ellis's commentary on Catullus, pp. xlv.-xlvii., and his introduction to the 64th poem.

him? In the teaching of Epicurus,[1] with its rigid rule of life, its strict checks on ambition, its stern repression of sensual desire, its insistence on the supreme duty of preserving tranquillity of soul, a man like Lucretius may have found a true deliverance, a deliverance of lower type, but as real as many a Greek youth had found in the words of Socrates. How often do the words 'peace,' 'perfect peace,' pacem, placidam pacem, recur in Lucretius's poem. To a man tormented by his own eager ambition and its disappointments, or deeply conscious of the tyranny of his own lower nature, there is no blessing greater than peace. Epicurus, indeed, insisted on the checking of all violent desires as the one thing needful. By this, to use his own phrase, 'all the storm of the soul is put an end to.'[2] It may be that the peace which Lucretius had found, and which his master had brought him, was to some extent ignoble and not much worth having. Epicurus, indeed, determined to live in the cellarage of the house of life, instead of the upper chambers, because it is safer, and there are such things as thunder-storms. Still Epicurus seems, judging from the poet's own words, to have brought to Lucretius *exactly that which he most needed*, and possibly at some crisis of his personal history. For this Lucretius was rightly grateful.

There is something pathetic in Lucretius's deep affection for Epicurus. Such warmth of gratitude is the sure sign of a noble nature.—The earnest way, too, in which he casts himself on the teaching of his master, sincerely and entirely accepting it, is not this the true attitude in which a man ought to receive all highest truth? Such, according to the parable, was the temper of the man who, having found a treasure, sold all that

[1] Epicureanism ought, as Martha says ('Poème de Lucrèce,' p. 35), to be called 'the doctrine of renunciation' rather than the doctrine of pleasure. Epicurus was himself a man of the most self-denying and almost ascetic type.

[2] λύεται πᾶς ὁ τῆς ψυχῆς χείμων (from the letter to Menoeceus, Diog. L., x.).

he had to purchase the field in which it lay. Natures like this are not common. Had Lucretius lived in a better day, and had the Master who was greater than Epicurus, or even than Socrates, called on him to follow, would not this ardent soul have cast aside all to follow him? and if so, he who is so entirely loyal to the truth he grasps, and who so utterly loathes the cowardice of superstition,—would he not to a higher truth than his old philosophy have proved faithful even to martyrdom? The test of any man's character is not merely the side on which he stands, but the manner in which he stands by it,—whether he be of the temper to 'contend to the uttermost,' or not. To Lucretius no one can impute the crowning condemnation of every frustrate spirit,—

> The unlit lamp and the ungirt loin.

So great earnestness and courage cannot exist only to be thrown away.

AN EXAMINATION OF M. GUYAU'S CHAPTER ON ATOMIC DECLINATION ("LA MORALE D'ÉPICURE." 1881).[1]

[1] This chapter is reprinted, with a few additions, from the "Journal of Philology," vol. xi., 1882.

ONE of the most valuable contributions to the history of ancient philosophy made in recent years is M. Guyau's brilliant work entitled 'La Morale d'Épicure' (2nd edition, Paris, 1881). This work, first published in 1878, was at once welcomed as important and eminently fresh in its treatment.[1] M. Guyau has devoted the whole of a masterly and admirably-written chapter (pp. 71-102) entitled 'Contingency in Nature the condition of Free-will in Man' to a study of the remarkable doctrine of Atomic Declination by which Epicurus attempted to explain Free-will. This great Epicurean doctrine has been discussed by almost no writer previous to M. Guyau, who justifies the length at which he has treated it by pointing out its importance. He speaks of it, justly we believe, as '*the central and truly original point of the Epicurean system,* namely, the relation of Free-will to Atomic Declination;'[2] and again, 'It is with regard to this point in particular that Epicurus might truthfully claim to owe his philosophy to himself alone.'[3] This chapter, evidently considered by the author to be unquestionably an important contribution to a true understanding of Epicureanism,[4] is

[1] Comparatively few notices of M. Guyau's work have appeared in this country. We extract the following from a lengthy one in the 'Athenaeum.' 'This work of M. Guyau's is full of suggestiveness, originality and value, and is based on a complete and masterly appreciation of the data existing. . . . As a study in ancient philosophy it is in many respects worthy to take its place beside even such a work as M. Ravaisson's "Métaphysique d'Aristote." . . . Those interested in the history of moral philosophy would be ill advised to overlook it, and no one can read it without profit.' —'Athenaeum,' Aug. 30, 1879.

[2] p. 99 (note).

[3] p. 73. With reference to the anecdote in Diog. Laert., x. 13.

[4] See p. 7 (note).

the part of his work which we now propose to examine. M. Guyau's explanation of the subject is in several respects a novel one, and is especially so in regard to one point, viz., his account of Epicurus's teaching as to Chance and the very important part which M. Guyau supposes it to play in the Epicurean philosophy. According to him, Epicurus believed that the element of Chance which we see at work in the world every day is the manifestation and outcome of a principle of 'Spontaneity' existing in Nature. This 'Spontaneity' is the consequence of the power of Declination possessed by the atoms. Thus Epicurus conceived both Free-will in man and the element of Chance in the world around him to be the result of the same power of Atomic Declination in its twofold working. We shall first state M. Guyau's theory, which he develops in a very subtle way, and then attempt to examine it. If his explanation be correct, it works a strange transformation in the accepted notions of Epicurean doctrine, and Epicurus, who is generally held to be a hard and bare materialist, must have attributed to Nature powers which in some respects remind us of the Fairy tales of our childhood or of the wilder dreams of Pantheism.

Epicurus, says M. Guyau, after having combated the religious idea of Providence or Divine caprice, found himself confronted with the scientific idea of Necessity. Thus his main philosophic aim was to escape from the notion of gods interfering with Nature on the one hand and to steer clear of the doctrine of Fate on the other. 'It is better,' said Epicurus, 'to believe in the fables of the gods than to be a slave to the fate of the natural philosophers. The myths allow us the hope of bending the gods by honouring them, but we cannot bend Necessity.'

'To imagine the gods above the world,' M. Guyau goes on, 'was to make oneself a slave: but to explain all things, oneself included, by necessary reasons which exclude our personal

Free-will, would be to do still more, it would be to suppress oneself. Absolute power of the gods or absolute power of the eternal laws, this is the alternative, while the impotence of man is the conclusion.' Epicurus was thus placed 'between the gods of Paganism and the Necessity of the Stoics or of the Natural Philosophers.' This was the dilemma which confronted him.

Epicurus was able to solve it only by adopting an entirely new philosophical position, taking his stand on which he was able to destroy Necessity and the power of the gods at the same time. 'To introduce into phenomena sufficient regularity that miracle may not be able to find place, and sufficient spontaneity that Necessity may no longer have any absolute, primitive, or decisive power—such is the double aim pursued by Epicurus.' How did he succeed in attaining it?

It is well known that Epicurus solved the difficulty, in a way satisfactory to himself, by assigning to the atoms the power of Declination. But for this power the world could never have come into existence, for otherwise the atoms could never have come into contact and produced the earth or the life upon it. It is the same power of spontaneous movement in the atoms of the soul which alone originates and renders possible the Free-will of man. Thus Epicurus had solved the difficulty. 'If all beings had within themselves naturally, instead of borrowing it from without, a spontaneous power whence their own movements should originate, might one not thus escape from the universal enchainment of cause and effect? Might not nature be conceived to be, essentially, at the same time without the gods and without Necessity?' Thus 'Democritus and Epicurus are as logical the one as the other: the first, admitting Necessity everywhere in the world, placed it in man also; the second, admitting Free-will in man, saw himself compelled to introduce an element of contingency into the world too.'

'It is commonly thought,' M. Guyau continues, 'that Contingency, placed by Epicurus at the origin of things, existed, according to him, at the origin alone and then disappeared in order again to leave room for Necessity. The world once made, the machine once constructed, why should it not go on by itself without any need of henceforth invoking any other force than Necessity?' The chain of destiny has been broken once, but closes again ring upon ring and clasps the universe afresh. 'According to this hypothesis Epicurus must have introduced declination into nature only as a kind of dialectic expedient and immediately made haste to withdraw it.'

This conclusion has been drawn from Lucretius's often-repeated statements that phenomena take place according to fixed conditions, and in particular that men, animals, trees are produced each after their kind from different germs, developing according to fixed methods. No organism can be produced at haphazard, without its proper germ and necessary conditions, for, says M. Guyau, translating Lucretius, 'each being is produced from fixed germs which are the object of scientific *certainty*' (seminibus quia certis quidque creatur). M. Guyau refers specially to the use of *certus* with reference to organic life, and continues, 'It is on this use of the word certus, several times repeated in reference to the germs of organisms, that the conclusion has been based that in the Epicurean system an unalterable fixedness of effects succeeds the freedom of the first cause, that " [after the world has once been formed] " this vast universe obeys and will eternally obey the laws of necessity, and that henceforth declination is incapable to break the enchainment of causes.' [1]

Such a conclusion however runs beyond the thought of Lucretius. 'Would certain philosophers of our own day who, like Epicurus, admit—rightly or wrongly—contingency in the

[1] The confusion of ideas implied in this paragraph is remarkable. See below, § 2, p. 221. (Of course, certus means much more than 'what is known with scientific certainty.')

universe, believe on that account that an apple-tree may produce an orange, or an orange-tree an apple ?' . . . 'It is one thing to believe that the universe, in its first principles, is not submitted to an absolute necessity, and another thing to believe in the sudden derangement of all natural laws or results. The spontaneous and initial movement cannot be calculated and determined beforehand (nec ratione[1] loci certa), but the combinations of movements once produced can be calculated and determined ; they constitute a fixed material which things require in order to come into existence (materies *certa* rebus gignundis).' It is not true that Epicurus supposes Declination to disappear from the world after it has been formed and henceforth to cease to exist in it. He holds the very reverse of this. 'Wherever the Epicureans speak of Declination, they consider it not as ended and done with, not as mere accident, a fortuitous exception to the order of things occurring once and never to be reproduced, but as a very real power which both the atoms and the individuals formed from the union of these atoms still retain.'

Man calls this power into use every day, nor does it exist in man alone, but in all forms of matter. M. Guyau quotes the famous passage on declination as the origin of our Free-will, and continues, 'Another passage relating not now to the declination of souls but to that of heavy bodies (non plus à la déclinaison des âmes mais à celle des corps pesants) is no less decisive. Evidently, says Lucretius, the heavy bodies which we see falling do not in their descent follow an oblique direction, but "who could distinguish that they absolutely to no extent decline from the perpendicular,"

sed nihil omnino recta regione viai
declinare, quis est qui possit cernere sese ?

Thus, following this somewhat simple conception of Epicurus,

[1] Here and at p. 78 M. Guyau reads by some strange mistake ratione instead of regione in the line—

nec regione loci certa nec tempore certo. ii. 293.

even before our eyes, even in the coarsest aggregations of matter, spontaneity might easily still retain a place ; it might manifest itself by an actual, though imperceptible movement, by a disturbance of which the effect will appear only after centuries.[1] *Everywhere then where the atom is found, in external objects as in ourselves, there will exist more or less latent the power of breaking necessity, and since, outside the atom, there is only void, nowhere will an absolute necessity reign : the Free-will which man possesses will exist everywhere, in inferior degrees, but always ready to awake and act.*'

'Can it be said that in placing Spontaneity everywhere, Epicurus placed everywhere a kind of miracle and thus returned without wishing it to the conception of a marvellous power like that of the gods? No! and Epicurus always thought himself able to reject the idea of miracle while at the same time defending the hypothesis of declination, which was dear to him. That there may really be miracle, two conditions must be realized; first, we must suppose powers existing outside of nature, then we must attribute to them a potency over nature large enough at once to modify, after a preconceived plan, an ensemble of phenomena. On the contrary, the spontaneity of the atoms is a power placed in the things themselves, not outside them, and, at the same time, this power is exercised only over a single movement; it oversteps the necessary laws of mechanics (ulterior and derivative laws) only on a single point and in a quite imperceptible manner. Spontaneous movements

[1] Assuming 'Spontaneity' to be a fact, we take leave to question whether the result of its working would, in consistence with the Epicurean doctrine as to the action of Declination, always be so imperceptibly small and slow as M. Guyau supposes. It is a principle of Mechanics that a very slight force may let loose a very great one, just as the huge boulder poised on a mountain-ledge may be finally cast down by some tiny rush of water. The spontaneous movement of a mass of matter, however slight, might still be able to give the initial impulse required to let loose a mighty force. Thus 'Spontaneity' might easily produce important results in Nature.

can have results only at length, by accumulation, by permitting new combinations, by thus aiding the march of things instead of hindering it: *spontaneity, if it exists, works to the same purpose as nature;* to believe Epicurus, we do not really disturb the laws of nature when, by a decision of the will impossible to determine, we resolve in such or such a way or take such and such a direction. Miracle, on the contrary, is in direct and formal opposition to nature: it is a violent arrest of the march of things. . . . *Spontaneity, on the contrary, precedes, follows and completes nature, hinders it from being a pure mechanism incapable of improvement:* it is for this that Epicurus maintains it: he hopes, rightly or wrongly, thus to counterbalance necessity, yet without disturbing the order of things' (pp. 91, 92).

Fragments of Epicurus's own writings and the statements of ancient writers show, says M. Guyau, that Epicurus believed 'Chance' or 'Fortune' to play a very important part in the world. 'Those external events which are not originally submitted to a necessary law, but to spontaneous causes the effects of which we cannot foresee, are referred to Chance.' Epicurus believed this principle of Chance or Accident which we see at work every day around us, to be the manifestation and outcome of the power of Spontaneity which resides in Matter.

'Chance does not mean for Epicurus the absence of cause: for we know nothing is done without cause, nothing comes from nothing: it is on this very principle that Epicurus rests in order to induce our Free-will on nature. Nor yet is Chance, as has been often said, the same as Free-will: for Epicurus always places the two terms, chance and liberty, parallel, without confounding the one with the other (ἂ μὲν ἀπὸ τύχης, ἂ δὲ παρ' ἡμᾶς). Chance in fact is exterior, liberty is interior.' 'Chance is a manner in which things appear in their relation to us: it is the unforeseen, the undeterminable, which occurs at an uncertain time and place. *But this element of the unforeseen is the result of a cause which hides itself behind Chance. This cause . . . is, in fact,* as we have seen, *the spontaneity of motion inherent in the atoms.* Chance is only the form under which this spontaneity reveals itself to us.' This, says M. Guyau, completely explains the passage of Plutarch which we can now better understand, 'Epicurus assigns the power of declination to the atoms . . . in order that Chance (τύχη) may be produced and that Free-will (τὸ ἐφ' ἡμῖν) may not be destroyed.' 'Τύχη *and* τὸ ἐφ' ἡμῖν *are the two modes of a Spontaneity identical at bottom,* to which Epicurus has just told us[1] that the destiny of the natural philosophers is reduced.'

All this has a practical bearing on man's life in the world. This external Chance when once manifested becomes a power more or less hostile to us,—Fortune. 'Fortune, it is true, is no longer a power absolutely invariable and unconquerable as destiny was. With changing and variable Chance, hope is always permitted, nay more, always enjoined. . . . Since no inflexible destiny can now impose itself upon us either without or within, nature cannot have dominion over us;—we on the contrary ought to command her by our Will. The wise man who might have been reduced to despair and helplessness before the Absolute of necessity or of divine caprice will recover all his strength when confronted with Chance, that is to say, at bottom, with spontaneity, that is, with a power which is no longer terrible like the unknown, but which he knows, nay more, which he carries within himself. He will then stand up like a wrestler against Chance and will struggle with it hand to hand;—a noble contest in which the wise man sure of his superior liberty is sure of his final triumph.' Thus, according to M. Guyau, in the struggle of man with nature, seeing that man has a high degree of Spontaneity and also Life

[1] Epicurus merely says (Diog. L., x. 133) that instead of Necessity being the mistress of all things, events are in reality due partly to Chance and partly to our own Free-will.

and consciousness, while he fights against things which possess Spontaneity only without life, man has an enormous advantage.

M. Guyau has now conducted us to the moral bearing of the question. He has shown how the Epicurean wise man need not tremble at Fortune with her turning wheel. 'Fortune or Chance has so little empire over the wise man that it is better,' said Epicurus, 'to be unfortunate according to reason than to be fortunate without reason.' In conclusion, he points out (pp. 99-102) 'the close solidarity established between man and the world which the doctrine implies.' 'Nature and man,' as he has before said, 'are so *solidaires* that we cannot find anything absolutely new in the one which should be wanting in the other: if we wish to recognize a principle of Spontaneity and liberty in ourselves, do not let us entirely withdraw it from things. We cannot set limits to Necessity and say, It reigns all around us, but it does not reign over us.' 'We naturally imagine that the whole universe may be subjected to Fate, without our Free-will, if it does exist, receiving any prejudice from it. But then, asks Epicurus, whence could this Free-will come? unde est haec, fatis avolsa, potestas? how could it be born and subsist in a world absolutely under the sway of necessary laws? . . . No, all causes are natural, and since "nothing comes from nothing," our Free-will comes from nature itself. It is curious to see Lucretius thus invoking in favour of Spontaneous Declination the famous axiom ex nihilo nihil, which has so often been urged precisely against this hypothesis.' According to Lucretius, what is in the effect exists already in the causes: if we can move at will, 'all the parts of our being which, by gathering together, have formed us, must possess an analogous power, more or less extensive, more or less conscious [! cf. Lucr. ii. 972, primordia . . . haut ullo praedita sensu],[1] but real.'

[1] Or ii. 990:—Seminibus . . . carentibus undique sensu.

'The adversaries of Epicurus attempted, as we have seen,[1] to escape from the dilemma which he laid down for them—either spontaneity in things or necessity in the soul, but it is doubtful whether they succeeded. In our own day the same dilemma still meets us. . . . Let there be a single being, a single molecule, a single atom in the universe in which spontaneity does not exist, and beyond doubt Free-will will no longer be able to find place within us: all existing things are *solidaire.* Inversely, if Free-will exists in man, it cannot be absolutely foreign to nature.' 'Hypothesis for hypothesis, we a hundred times prefer the Epicurean *clinamen* to the vulgar doctrine of Free-will restricted to man.'

M. Guyau does not examine how far 'this universal spontaneity, this element of variability introduced into the universe, may agree with the theories of modern science as to the equivalence of forces and the mechanical laws of evolution.' His task has been 'simply to look for the true meaning and to show the historical importance of one of the chief theories of Epicurus.'

Most students of ancient philosophy will be astonished at the entirely new light which this chapter of M. Guyau's pours over Epicureanism. So reasonable and consistent with the logical results of some part of Epicurean doctrine is his explanation, so forcibly does he grasp and express it, and so skilfully does he handle and combine the evidence which seems to support his opinion, that we seem at first compelled to admit its

[1] M. Guyau refers to the ingenious argument of Carneades, who taught that declination was unnecessary, since both the atoms and man have power to move without any external cause in virtue of their own *nature.* (ipsius individui hanc esse naturam, ut pondere et gravitate moveatur, eamque ipsam esse causam cur ita feratur . . . similiter ad animorum motus voluntarios non est requirenda externa causa; motus enim voluntarius eam naturam in se ipse continet ut sit in nostra potestate, nobisque pareat, nec id sine causa, eius enim rei causa ipsa natura est.—Cicero, 'De Fato,' xi.)

historic accuracy. And if so, must not Epicureanism be the very reverse of what it has been thought? how much of the marvellous it must have included? If 'Spontaneity' exist even in brute-masses of matter, if the stone which I hold in my hand—not merely its individual atoms, as Epicurus did indubitably assert—but if the mass of stone itself possess 'Spontaneity' and Will so that it can move in any direction at pleasure, what matter though its movements be so slight as to be imperceptible to the human eye,—does not this remind us of those Fairy-tales, which show how in simpler ages than this men found it easy to credit all sorts of magical powers in Matter, and looked upon all objects of the outer world as animated with a life resembling their own?[1] This tendency is seen in such stories as that of the rocks which the early Greek mariners believed had the habit of dashing together so as to crush unwary ships, of the good ship Argo which has sunk so deep into the sand that she cannot be launched, but when the prophet sings to his lyre, she rises out of her sandy bed and rushes forward into the sea, of the fairy's bean which in one night grows into a beanstalk higher than the tallest tree, of the automatic cudgel which can beat a man at its owner's bidding, or a hundred others. We are even reminded of Hans Andersen's delicious stories, where everything in the world, from the Fir-tree, the Rose-bush and the Daisy down to the Old Lamp and the Silver Shilling, possesses personality and consciousness each

[1] Comte has described this under the name of 'Fetichism' as a necessary stage of human development. It is the tendency of man, as seen in the history of every race, to look upon the world around him as animated like himself in greater or less degree. Comte's language on this subject strikingly reminds us of M. Guyau's description of 'Spontaneity,'—'pur fétichisme constamment charactérisé par l'essor libre et direct de tendance primitive à concevoir tous les corps extérieurs quelconques, naturels ou artificiels, comme animés d'une vie essentiellement analogue à la nôtre avec des simples différences mutuelles d'intensité' ('Philosophie Positive,' vol. v., p. 30).

after its own degrée and kind. True, Epicurus asserted for his atoms and, according to M. Guyau, also for masses of matter in every form, *Will* and consequent power of motion *without Life* and consciousness. But the common mind is utterly incapable of drawing such a distinction, and where Will is, it must without fail conceive Life and all its attributes to be also. 'Everywhere where the atom is, in external objects as well as in ourselves, will exist more or less latent the power of breaking necessity.' . . . 'The Free-will which man possesses will exist everywhere in inferior degrees, but always ready to awake and act.' . . . The atoms which have formed our bodies must possess a power of Free-will 'analogous to our own, more or less extensive, more or less conscious, but real.' And if this 'Spontaneity' residing in what we call dead Matter, has such power as to produce the fortuitous and unexpected in Circumstance, that which we cannot calculate upon and which happens at times and in places where we do not look for it, either coming to baffle us or bringing us success, so that what we call 'Chance' in the affairs of daily life is the direct result of the long-continued blind-working of 'Spontaneity' in Matter, does not a conception like this bring us nearer to the world of Fairy-tale than to that of Science, still less to that of Materialism ?—What strange results might come of such a potency in Matter ! One cannot help thinking how a power like this, were it possible for it to exist in a world such as ours and under the domain of natural law, would in many ways render Nature far more terrible to man than she is. How easily might such a force set the avalanche sliding on the mountain-side, or bring down the hanging rock upon the passer-by, or set the tempest brewing ! And 'Spontaneity' would be the more dangerous since, unlike the other forces of Nature, it has no fixed methods but manifests itself incerto tempore ferme Incertisque locis, so that we cannot forecast its working.

But it is now full time to examine M. Guyau's evidence.

Is his explanation of this important Epicurean doctrine historically accurate, or is it not?

1. How does M. Guyau reconcile the existence of 'Spontaneity' in things with the leading Epicurean principle of the constancy of natural laws, a principle which we have shown was grasped as strongly by Lucretius as it is by any modern man of science?

2. In the first place, M. Guyau appears to us never fully to realize or give account to Epicurus's distinct and decided grasp of the fact of Law in Nature. Indeed he appears actually to contradict it. He objects (pp. 87-9) to our supposing that according to Epicurus, 'contingency existed solely at the origin of things and afterwards disappeared in order again to make way for necessity,' and that 'this universe now obeys and will obey eternally the laws of necessity, and that declination is henceforth unable to break the enchainment of causes.' This part of M. Guyau's chapter involves a rather intricate confusion of ideas, and is in one respect entirely false. According to Lucretius and his Master, Law reigns everywhere in Nature and 'Necessity' is a name given by both[1] to the order of Nature resulting from natural law, though Lucretius[2] uses the

[1] Epicureans would probably have assigned the movements of the heavenly bodies as the readiest instance of that which has 'Necessity' for its cause. Epicurus does so in his letters (Diogenes Laertius, x. 77 and 113), and similarly he speaks of Necessity (*ἀνάγκη*) as a possible First Cause of the movement of the heaven or of the stars (ibid., x. 92. *κατὰ τὴν ἐξ ἀρχῆς ἐν τῇ τοῦ κόσμου γενέσει ἀνάγκην ἀπογεννηθεῖσαν*; cf. also x. 93). We may compare *καὶ τίσιν ἀνάγκαις ἕκαστα γίγνεται τῶν οὐρανίων* ('Memorabilia,' i. 1, 12; cf. i. 1, 15), where Xenophon uses the word in the precise meaning of 'natural laws.' 'In the same way Aristotle says that 'great storms and floods recur *διὰ χρόνων εἱμαρμένων* (*i.e.* at fixed periods), just as winter occurs at a given season of the year' ('Meteor.,' i. 14).

[2] As at v. 309-10,

> nec sanctum numen *fati* protollere *finis*
> posse neque adversus naturae foedera niti,

where fati finis, 'the limits of fate,' refers to the same thing as naturae word in this sense comparatively seldom. Lucretius firmly believes that nowhere in Nature can you escape from law. In this sense Epicureans *did* conceive the world after its origin 'to obey the laws of necessity,' to be 'subject to an absolute necessity.'

3. M. Guyau has referred to those passages in which the word *certus* occurs, and frequently with reference to the germs of organisms, *e.g.*

> seminibus quia certis quaeque creantur. i. 169.
>
> atque hac re nequeunt ex omnibus omnia gigni,
> quod certis in rebus inest secreta facultas. i. 172.
>
> omnia quando
> paulatim crescunt, ut par est, semine certo. i. 189.
>
> si non materies quia rebus reddita certast
> gignundis[1] e qua constat quid possit oriri. i. 203.

foedera. So the famous passage on Free-will, if correctly understood, distinctly implies that the world outside man is absolutely governed by fate (cf. the context of ii. 254, fati foedera; 257, fatis avolsa potestas), and here evidently Lucretius shows that he conceives the laws of nature as fate.—Occasionally Lucretius uses vis in the sense of 'necessity' instead of fatum or necessum, as at ii. 289, where externa vis is opposed to necessum intestinum. The passage vi. 29-32, which touches on the source of evil in human affairs,

> quod fieret *naturali* varieque volaret
> *seu casu seu vi*, quod sic natura parasset

seems to mean that it is indifferent whether you call the cause of evil from one point of view 'natural chance,' seeing that, as concerns us, it is not fixed or decreed whom it is to injure, or from another standpoint 'natural necessity,' since if we come into collision with it, it will and must according to nature's law inevitably injure or crush us. The passage of course implies that you must *not* ascribe evils either to Divine Providence or to Fate. Here vis naturalis certainly refers to the 'necessity' which is the consequence of natural laws. See also Munro's note on v. 77, and on vi. 31.

[1] At p. 89 M. Guyau seems to misunderstand this. He renders 'une matière certaine dont les choses ont besoin pour naître.' But materies certa refers to the atoms and their unchanging character. Similarly at p. 69 when M. Guyau translates finita potestas denique cuique Quanam

He might have added many other passages, such as—

omnia quando
seminibus certis certa genetrice creata
conservare genus crescentia posse videmus.
scilicet id *certa* fieri *ratione* necessust. ii. 707-10.

certum ac dispositumst ubi quicquit crescat et insit. iii. 787.

Cf. also v. 669-79 ; v. 923-4 ; v. 1436-9.

M. Guyau (pp. 88-9) appears to us considerably to misunderstand the force of certus in these passages. It refers to the *fixity* and *unchangeableness* of law as manifested in natural productions. Things which are entirely subject to natural law, such as the growth of trees and plants, and the development of living bodies, animals and men, each after its kind and from its own proper germ, are 'fixed' (certus) in respect of the time, place and conditions of their coming into being and continuing in existence. On the other hand, the will of man is not thus predetermined by causes outside himself ; it acts

nec tempore certo
nec regione loci certa.

M. Guyau does not by any means sufficiently distinguish between the two Epicurean principles of absolute fixity of law (sometimes in Epicurean language called 'Necessity') in Nature and perfect Spontaneity of Free-will action in man.

4. The question now very naturally occurs to us,—If Matter everywhere possesses 'Spontaneity' and is always exerting it, how can this be without interfering more or less with the constancy of natural law, the principle upon which all Epicurean science was based ? However slight and gradual such declination may be, if all bodies everywhere are exerting it, they must,

sit ratione, 'par quelle raison chaque chose *n'a qu'une puissance limitée*,' he misunderstands finita and misses the idea, which is the fixity and definiteness of natural laws. Cf. also p. 70.

inevitably, more or less disturb the orderly sequence of natural phenomena, if not destroy the conditions under which Law is possible. M. Guyau appears to think that the slightness of the amount of such action (une perturbation dont l'effet n'apparaîtra qu'après des siècles) will produce a variation so small and slow as not to interfere with nature, but as we have already pointed out, if we assume 'Spontaneity' to be possible, the amount of its action cannot be counted on. At one time, its working in some huge mass might be imperceptibly small, at another it might chance to be enough to let loose and set a-going some vast atomic machinery with far-reaching consequences, or if it chanced to combine with a series of other spontaneous movements in other bodies and from other sources, its results might be enormous and speedy enough.—In any case, however, if such a power be exerted by Matter, there can be no fixed laws of Nature, no *foedera certa*, no *terminus alte haerens*. A far less shrewd thinker than Epicurus could hardly have failed to see that 'Spontaneity' in the various forms of Matter cannot exist side by side with absolute laws of Nature.

5. M. Guyau has foreseen this and tries to guard against it by assuming that 'Spontaneity' cannot disturb natural order, because it ('va dans le sens de la nature') works in harmony with nature. The assumption is baseless, and rather a bold one. Why should not 'Spontaneity' as well work *against* natural order?

6. But supposing atoms to possess the power of movement in any direction at will, does it follow that any body formed out of atoms, say a mass of stone, can *as a body* possess the same power of movement which its atoms have *as atoms?* Certainly not, according to Epicurus's conception of the atoms: rather would one of its component atoms move in one direction, another in an opposite, and thus they would counteract each other, and the body remain inert. M. Guyau (quoted above at p. 213) states

that one passage (ii. 249-50) decisively shows that Lucretius believed in 'the declination of heavy bodies' as well as in 'the declination of minds.' But M. Guyau has entirely misunderstood the passage in question,—

> quare etiam atque etiam paulum inclinare necessest
> corpora; nec plus quam minimum,—*ne fingere motus*
> *obliquos videamur et id res vera refutet.*
> namque hoc in promptu manifestumque esse videmus,
> pondera, quantum in sest, non posse obliqua meare,
> ex supero cum praecipitant, quod cernere possis;
> *sed nil omnino recta regione viai*
> *declinare quis est qui possit cernere sese.* ii. 243-50.

We allow that this passage comes in such a connection as to be most easily misunderstood, and at first and even second reading it certainly appears to bear the meaning which M. Guyau has given it. But really it amounts to this,—'We never see falling bodies swerve, it is true,' says Lucretius, 'but that does not prove it to be against nature and impossible for such a thing to happen. The human eye is incapable of deciding that falling bodies move in an *absolutely straight* line. A stone falling to the ground may slant to an exceedingly small extent for all that we can tell. Therefore, *so far as the evidence of sense is concerned,* it is not impossible that the atom should swerve (nec plus quam minimum) to a very slight extent.' It is well known what stress Epicurus laid on the principle that the senses cannot deceive, and it is the apparent testimony of sense, of observed facts (res vera), which Lucretius is combating in these two lines.

7. This passage is M. Guyau's main evidence for the assertion that, according to Epicurean belief, masses of matter have the power to decline as well as atoms. What other proof does he bring forward? Out of all the authorities quoted, only one passage from Plutarch contains anything at all distinct enough to appear to support M. Guyau's theory, but so skilfully does he lead up to his conclusion that the evidence seems stronger

than it is. Indeed, the passages referred to bear only in the vaguest way on the present subject. They simply assert that Epicurus often attributed events to Fortune. But most ancient philosophers speak in the same way and assign more or less power to Fortune in ordering what comes to pass. M. Guyau quotes one passage of Plutarch, translating it as follows—'Epicurus assigns declination to the atom . . . in order that chance may be produced and free-will may not be destroyed:—ἄτομον παρεγκλῖναι (spontaneity of declination) . . . ὅπως τύχη παρεισέλθῃ (external chance which is the form of it) καὶ τὸ ἐφ' ἡμῖν μὴ ἀπόληται (inward liberty which is the feeling of it).' This commentary builds a good deal on Plutarch's incidental and sarcastic reference to Epicurus, even were the sentence exactly as M. Guyau has quoted it. Plutarch does not refer to Epicurus at all in the context, but simply makes a fling at him in passing, as follows: 'The philosophers do not allow Epicurus, even in order to account for the greatest things, to assume so small and unimportant a matter as the least possible declination of a single atom,[1] in order that the worlds and living creatures and Fortune may be smuggled in,[2] and that our Free-will may not be destroyed.' (In the next sentence Plutarch passes on to a quite different subject.) Instead of saying that Epicurus introduced the doctrine of Atomic Declination principally or solely to account for Chance, as M. Guyau's quotation would certainly make us suppose, Plutarch is stating correctly enough the general objects which Epicurus thought to effect by Declination, viz., to allow the origin of the worlds and of man, and to

[1] Plutarch, 'De Solertia Animalium,' c. vii. Probably the words ἄτομον παρεγκλῖναι μίαν are not intended to be understood literally in the sense that Epicurus required the declination of 'a single atom' only to begin with: but Plutarch's statements as to Epicureanism are not always strictly accurate, as in the passage ('De Plac. Phil.,' i. 7) where, in speaking of the fourth incorruptible element of the Epicureans, he certainly confuses it with the ὁμοιομερείαι of Anaxagoras.

[2] ὅπως ἄστρα καὶ ζῶα καὶ τύχη παρεισέλθῃ; (παρεισ' 'may slip in at the side,' used sarcastically).

render Free-will possible (*ὅπως . . . τύχη παρεισέλθῃ* meaning simply 'to get rid of Necessity').

8. There is no doubt that the Epicurean writers spoke much of Chance. In the Epicurean system, which rejected all and any Providence, Chance must from the very facts of human nature have come to be an important item in everyday calculations about human affairs. Epicureans refused to own any Divine agency in the world, but practically they had set up a new Divinity, Chance, which was for them a real enough one. Chance must have been often in the mouth of an Epicurean,[1] just as naturally as Providence was in that of a Stoic, or 'the hand of God' in that of a Puritan. It was simply natural that Lucretius should pray that the abstraction Fortuna gubernans might avert the end of the world.

Lucretius not merely opposed the notion of Gods from time to time interfering with nature, but he like other Epicureans would have combated with equal ardour the belief, held in a very noble form by the Stoics, in a universal Providence ordaining each and every event of human life as well as maintaining all the ongoings of nature. Such a conception would have appeared to him only another form of Necessity and almost equally objectionable. In human affairs Providence (according to the Pagan notion of it, as represented by Virgil's gods and goddesses, who bitterly persecute the human beings who have unwittingly and often innocently given them offence) had come to be dreaded. Chance seemed less formidable.

It is very difficult for us, accustomed to modern phraseology, to understand the exact meaning of such words as Chance and

[1] Cf. the opinion ascribed by Hippolytus ('Ref. haer.,' i. 22) to Epicurus,—*ὅλως πρόνοιαν μὴ εἶναι μηδὲ εἱμαρμένην ἀλλὰ πάντα κατὰ αὐτοματισμὸν γίνεσθαι*:—'There is neither Providence at all, nor yet Destiny, but all things take place by Chance' or 'happen of themselves.' As the Lucretian parallel for *πάντα κατὰ αὐτοματισμὸν γίνεσθαι*, we might quote

natura videtur
ipsa sua per se sponte omnia dis *agere* expers. ii. 1090-2.

Necessity in the Epicurean as also in other systems of ancient philosophy. For example, Stobaeus (i. 206) tells us that 'Epicurus distinguishes among Causes, that by Necessity, that by Free-will, and that by Fortune,' Ἐπίκουρος (προσδιαρθροῖ ταῖς αἰτίαις τὴν) κατ' ἀνάγκην, κατὰ προαίρεσιν, κατὰ τύχην.[1] Perhaps we may best explain this by taking an instance, such as the incident, used as an illustration by Lucretius, of the Roman admiral and his fleet destroyed by the tempest. Here there would be, according to the phraseology just quoted, three 'Causes' at work. (1) 'Necessity,' or as Lucretius once calls it *vis naturalis*, 'natural Necessity,' *i.e.*, the laws which produce storms and which cannot do otherwise than produce them at their given time and place,—certo tempore, certo spatio. At the present day we should call this, far more appropriately, Natural Law.[2] (2) Free-will, which works incerto tempore ferme Incertisque locis. The admiral was free to have taken another course or to have delayed his voyage till a safer time, but he chose to sail then and in the direction where the tempest was to burst. (3) Chance, that is to say, the way in which the forces of nature, in their working, bear on man. It

[1] Plutarch ('De Plac. Phil.,' i. 29) and Stobaeus (i. 218) both assert that Epicurus held Fortune to be ἀσύστατον αἰτίαν;—the expression may be Epicurus's own or not. Epicurus himself, in a letter preserved by Diogenes Laertius (x. 134), distinctly says that Fortune is neither θεὸν nor yet αἰτίαν. —M. Guyau attributes to Epicurus the saying found in Sextus Empiricus (p. 736, ed. Bekker, 1842), τῶν γινομένων τὰ μὲν κατ' ἀνάγκην γίνεται, τὰ δὲ κατὰ τύχην, τὰ δὲ παρ' ἡμᾶς. There is no ground whatever for assigning this to Epicurus on the authority of Sextus Empiricus, who does not in any way refer to him.

[2] See notes on § 2, pp. 221-2. Epicurus boasted that he had cast out Necessity from the moral world. Here he claimed to have substituted for it the two notions of Chance and Free-will (Diog. L., x. 133). He still called the laws of nature, in so far as they absolutely govern the world outside man, 'Necessity,' but in the physical world also the principle of Law which he had done so much to establish, was really destined to substitute for the notion of Necessity a higher Idea, though neither Epicurus nor Lucretius had any anticipation of this.

might easily have been otherwise. The storm might have raged either sooner or later or over another portion of the sea, but as it coincided with the course and the time which the admiral chose, nature could do nothing else than destroy him. Chance comes into play where the forces of nature come to bear for good or evil on human affairs.[1]—These three principles do not by any means stand in the same category. 'Necessity' and Free-will are both causes, but Fortune is in no sense a cause, and can only be called so by a popular and unscientific use of language.

9. While for the reasons given we cannot allow that M. Guyau's theory of 'Spontaneity' is correct, or that there is evidence to prove that Epicurus or any of his followers held such a doctrine, still it might be asserted with some reason that it is an entirely logical inference from the doctrine of Atomic Declination. Supposing the power of declining to exist in atoms, and that they exert it, if we endeavour by an effort of imagination to conceive the effect, would it not be something like 'Spontaneity' which might naturally enough manifest itself in the accidental and unforeseen of circumstance and of human affairs. But even though it were a logical deduction from one principal Epicurean doctrine, this would not be enough to prove it historically correct. It would merely prove Epicurus guilty of inconsistency. We certainly cannot agree with the remark which M. Guyau somewhere makes that 'in Epicureanism there are no inconsistencies, but only a few false deductions.'

10. It may very naturally be asked, How did Epicurus, who allowed so remarkable a power as Declination to exist in atoms, suppose it practically to disappear after these atoms have combined to form matter? He supposes it still to remain

[1] The words *ἀνάγκη* and *τὸ αὐτόματον* occur in somewhat strange collocation in the interesting fragment published by Gomperz ('Neue Bruchstücke Epikur's,' 1876, pp. 8-11), *ἐν τῷ τοῦ περιέχοντος καὶ ἐπεισιόντος κατὰ τὸ αὐτόματον ἀνάγκῃ.* Here Epicurus is evidently defending the freedom of the mental processes in reference to his theory of Perception by Images.

and work within them while confined in the various forms of matter, but how comes it to exert no farther influence? How does it work to such different effect in a rock and in a man? We have given one reason above (see § 6) which may partly explain this. It seems to us, so far as we can make out, that Epicurus assumed, whether reasonably or not, that the power of Declination, while still remaining and working in the atoms, would be virtually nullified by various counteracting causes,—by the conditions of the world which, when once it is formed, tend to hold things together[1] (the same forces which, when atoms have united in the manner necessary (concilium) to form any kind of substance (res), compel them to remain thus united and keep matter from dissolving into atoms), and partly, he would no doubt have said, by gravity, which would have a resisting influence.[2] Besides, Free-will is proportionally a far feebler power in gross matter, formed of coarse atoms (which are also heavier and harder to move), than it is in the soul, which is composed of exceedingly fine and smooth ones. Thus Free-will would exist in far less intensity in gross matter than in the soul, and be far more easily held in check. Such considerations must naturally have kept Epicurus from allowing that masses of matter can decline as the atoms can. For one thing, Lucretius is very conscious (and naturally enough) that an atomic chance-made world (such as he conceives ours to be) is exceedingly liable to destruction and may any day in a moment fall into ruins and pass away. It is curious how often he re-

[1] To a certain extent Lucretius conceives the plagae extrinsecus undique (i. 1042) or ictus externi (i. 1055) to act in this way. The atoms not combined in matter form an ever-tossing ocean, which is constantly beating against the surface of every object. These continual shocks produce a pressure from without which tends to hold things together and to keep the world in existence—

summam
conservare omnem quaecumque est conciliata. i. 1042-3.

[2] We may contrast the influence assigned to gravity at ii. 288-9.

minds us of the many possible causes which might bring this about. Would it not have appeared to him that the existence of such a power as 'the declination of heavy bodies' would render it impossible for a world, formed like ours, to hold together for a day? But certainly a thinker so shrewd as Epicurus could hardly have conceived such a power to exist in bodies *without also seeing that this would interfere more or less with the regularity of Law in Nature,* a fact which he so firmly and thoroughly grasped.

M. Guyau has made a vigorous endeavour to grasp the whole subject from every side and from all possible points of view,[1] and even while disagreeing with him, few will read this very remarkable chapter without feeling that he has flashed light round him. Aided by his wide knowledge of both ancient and modern philosophy, he makes us vividly realize the philosophical difficulty which Epicurus had to encounter, and also the solution offered by him. If M. Guyau's explanation of one leading Epicurean theory has been shown to be in part unsatisfactory and unfounded, it must not be supposed that the rest of his book is open to similar charges with the chapter which we have been examining. This would be unjust indeed. In reality, the picture of Epicurus as a teacher which M. Guyau's work gives us is drawn not merely with wonderful literary skill but with equal vigour and grasp of philosophic penetration. Certainly Epicureanism owes much to French scholarship, from Gassendi to M. Martha, whose 'Étude sur Lucrèce' is an admirable study of the 'De Rerum Natura' in its poetic and moral aspects, and finally to M. Guyau. Both the latter writers have done much to redeem the philosophy of Epicurus from the base estate in which it has so long lain, hopelessly covered and hid

[1] Notably, however, M. Guyau omits to touch in any way on the subtle adaptation between Atomic Declination and Epicurean psychology, a subject which we have attempted to explain elsewhere. ('British Quarterly,' April, 1882, p. 329 ff.)

with abuse and misrepresentations in many respects utterly undeserved.

It is needless to point out the close relation of M. Guyau's doctrine of Spontaneity-in-things to the philosophy of Schopenhauer. Guyau conceives 'Spontaneity,' and Schopenhauer 'Will,' to exist in Matter under its every form, attended by a greater or less degree of consciousness. The German philosopher sees in Will the real essence of the inorganic world as well as of vegetable and of animal life. He applies to the operation of natural forces such as heat, gravity, electricity, words which, consistently with his doctrine, are specially sought out from the vocabulary describing the efforts of human beings in whom Will takes its highest form. 'When we attentively consider with what irresistible *striving* water hurls itself into a hollow, the *perseverance* with which the magnet turns to the north, the *ardent desire* of the iron to cling to the magnet, the *violence* with which the two opposite poles of electricity seek to rejoin; when we observe with what rapidity, with what regularity of shape, with what determined *effort* in fixed directions the crystal forms; when we reflect with what *elective choice* bodies in the fluid state *seek* and *fly* from each other, *unite* and *part*; when we find, in short, within ourselves a burden as it were, the *striving* of which towards the terrestrial mass drags down our body, . . . we shall require no great effort of imagination in order to recognize that that which, in us, follows a fixed end in the light of intelligence, and that which, in the world, is but a blind, deaf, limited, invariable tendency, is one and the same thing—almost as the dawn and the full noon are both due to the rays of the sun—and that this thing is will. . . . The objectivation of will, becoming gradually more distinct, manifests itself in the vegetable world. . . . One plant wills a moist situation, another a dry, another a lofty one; one strives towards the light, the other towards the water. The climbing plant seeks a support; the tree cracks rocks, or bursts a wall by the persistent effort which it makes to develop itself,—and so on. All which things are due to the inferior form of will, which Schopenhauer calls excitation.'[1] We quote at such length in order to show how other philosophers, besides M. Guyau, have found in Nature the manifestations of Will. Not that the two doctrines coincide, for Schopenhauer finds in the inorganic world nothing but cause, working in a fixed order.

Although the theory of 'Spontaneity' so ingeniously set forth by M. Guyau was no part of Epicurean belief, and is thus historically incorrect, it is still interesting and a memorable one.[2] After the notion of Spontaneity working in the material substances everywhere around us, and having power to produce all that in daily life we call Chance or accident, has entered into the mind, it is a thought which, however unreasonable in some respects, one cannot help recurring to. Whatever distant suggestion of truth it may contain, we instinctively reflect that Nature is terrible enough and the world hard enough for man without the interference of a blind uncontrollable power like this, whose laws we could never hope to master, and which would ever and again transform the regular order of Nature into a mere 'Come what will.' Still M. Guyau's theory has even a certain philosophical value—and, whether in its supposed connection with Epicureanism or for its own sake, it will, we believe, from time to time be returned to and discussed afresh.

[1] This abstract, containing almost the words of Schopenhauer, is translated from M. Ribot's 'Philosophie de Schopenhauer,' Paris, 1874, pp. 76-8.

[2] Does not the notion of 'Spontaneity' in things remind us a little, of course merely in certain aspects, of Goethe's 'Daemonic Principle'? In Goethe's own words, 'the Daemonic is that which cannot be explained by reason and understanding.' It 'resembles Chance, for it evolves no consequences.' It manifests itself not in man merely, but 'in all corporeal and incorporeal things.' 'It is particularly perceptible in events, and indeed in all which we cannot explain by reason and understanding.'

APPENDIX.

APPENDIX.

I. *Page 2, note. Manufactured Articles.*

THIS phrase has been much misunderstood. In a letter to Bishop Ellicott, Professor Clerk-Maxwell explains it as follows:—

'The comparison of atoms or of molecules to manufactured articles was first made by Sir J. F. D. Herschel. . . . The comparison was criticised (I think in a letter to "Nature") by Mr. C. J. Munro, and the latter part of the *Encyc. Britannica* article, "Atom," is intended to meet this criticism, which points out that in some cases the uniformity among manufactured articles is evidence of want of power in the manufacturer to adapt each article to its special use' ('Life of James Clerk-Maxwell,' by Professor Lewis Campbell, 1882, p. 393).

In the paper on Atoms here referred to, Clerk-Maxwell points out that the 'uniformity of manufactured articles may be traced to very different motives on the part of the manufacturer.' In certain cases it is due to economy of production. In another class it is uniformity which constitutes the special value of the articles, as in the case of a Whitworth bolt (these being made in given and precise sizes), or of a printed book. 'In the third class not a part only, but the whole of the value of the object arises from the exact conformity to a given standard. Weights and measures belong to this class, and the existence of many well-adjusted material standards of weight and measure in any country furnishes evidence of the existence of a system of law, regulating the transactions of the inhabitants, and enjoining in all professed measures a conformity to the national standard.' It is the characteristic of the last class, viz., 'quantitative accuracy,' which induced Maxwell to compare the atoms to manufactured articles. 'What I thought of,' he says in the same letter to Bishop Ellicott, 'was not so much that uniformity of result which is due to uniformity in the process of formation, as a uniformity intended and accomplished by the same wisdom and power of which uniformity, accuracy, symmetry, consistency and continuity of plan are as important attributes as the contrivance of the special utility of each individual thing.'

II. *Page* 10. *Laws of Nature and Divine Premovement.*

IN a pamphlet recently reprinted ('Science, Prayer, Freewill and Miracles,' Burns and Oates, 1881), an able Roman Catholic writer (Dr. W. G. Ward) has effectively criticised the reasoning of Tyndall and others. Dr. Ward's conception of a 'Divine premovement' of events is by no means a novel one, though it has never before been worked out with so much force and grasp. He has recourse to a somewhat grotesque illustration, which however helps us to realize the question vividly enough. He imagines some mice, endowed with human or quasi-human intelligence, to be shut up in a musical instrument like a piano, but 'immeasurably more vast in size and more complex in machinery.' In this instrument the intermediate links between the player's premovement on the one hand and the resulting sound on the other, are not two only, as in a piano, but two hundred. On the polychordon some one is unintermittently playing, but playing on it just what airs may strike his fancy at the moment. The mice hear the music, and philosophize as to its origin. Successive generations of philosophical mice have actually traced one hundred and fifty of the two hundred phenomenal sequences through whose fixed and invariable laws the sound is produced. The colony of mice, shut up within, are delighted with the success which has crowned the labours of their leading thinkers, and the most eminent of these addresses an assembly as follows: 'We have long known that the laws of our musical universe are immutably fixed; but we have now discovered a far larger number of those laws than our ancestors could have imagined *capable* of discovery. Let us redouble our efforts. I fully expect that our grandchildren will be able to predict as accurately, for an indefinitely preceding period, the succession of melodies with which we are to be delighted, as we now predict the hours of sunrise and sunset. One thing, at all events, is now absolutely incontrovertible. As to the notion of there being some agency *external* to the polychordon,—intervening with arbitrary and capricious will to produce the sounds we experience,—this is a long-exploded superstition, a mere dream and dotage of the past. The progress of science has put it on one side, and never again can it return to disturb our philosophical progress.' The meaning of Dr. Ward's parable is clear. Two hundred absolutely fixed laws intervene between the player's premovement and the resulting sound, but this fact does not tend ever so remotely to show that there is not an intelligent player, or that his premovement is not absolutely unremitting. In like manner, though scientific men have discovered that the laws of nature are absolutely fixed, and though we have already mastered many of them, this '*would not tend ever so remotely to show, that those laws are not at each moment directed to this purpose or to that by an immediate and uncontrolled Divine Premovement.* God's real ends cannot be more inscrutable to us . . . . than would be the ends of a human performer to the mice within this supposed polychordon. . . . And as a player on the polychordon may readily be induced, at the smallest request of a little child, to produce this particular musical result rather than some other, so the heartfelt prayer of the humblest Christian may powerfully affect God's premovement of the physical world.' Dr. Ward's illustration of the philosophical mice brings the question home and makes it palpable in a way that excuses its grotesqueness.—He further draws a distinction, important for this subject, between *cosmical* phenomena, such as the hours of sunrise and sunset, or the periodical return of comets and eclipses, which are produced by an incredibly vast machinery, in which this earth plays a very subordinate part, and earthly phenomena such as the weather, the violence of the wind, and disease, which are due in great measure to agencies acting exclusively within the region of our planet. The course of cosmical phenomena is steady and amenable to calculation, while the course of earthly phenomena is variable and incalculable. Prayer has to do with the latter exclusively. 'It is most remarkable, and bears thinking of again and again, that the only power of indefinite prediction which science has ever procured, concerns cosmical phenomena and not earthly.' Again, if God premoves earthly phenomena, why does He will that their causation should be so complex? Dr. Ward replies most forcibly, 'It is not the general law of God's Providence that the truths of religion *shall* be visible and palpable facts; but, on the contrary, that they shall give occasion to the merit of faith. Let it be assumed, then, that God does premove earthly phenomena; and let the further very obvious supposition be also made, that He does not desire this premovement to be a visible and palpable fact. On this supposition, He would act just as we maintain that He has acted. *He would make earthly phenomena to proceed on so complex a chain of causation, that His assiduous premovement of them eludes direct observation.*'

Dr. Ward's theory of a Divine premovement is admirably stated, but, as we have elsewhere hinted (see page 95), have not all really spiritual natures worked out the problem much in this way, namely, that God is *always* at work, *everywhere* in Nature? To certain minds the existence of fixed laws which produce results, seemingly of themselves, forms an insurmountable barrier to the recognition of God. Yet even though He work beyond so vast and intricate a wheelwork of natural laws, a man like Socrates, and others who are courageous enough and self-denying enough to follow with humility the method of Socrates, the χαλεπὴν καὶ μακρὰν ὁδὸν, have never failed to realize, with more or less completeness, the Divine hand behind all the intermediate machinery.

III. *Page* 14. *Different shapes of the atoms.*

IN the following passage Theophrastus, a contemporary of Epicurus, describes the different forms of the atoms. He refers to the atomic theory of Democritus.

Δημόκριτος δὲ σχῆμα περιτιθεὶς ἑκάστῳ, γλυκὺν μὲν τὸν στρογγύλον καὶ εὐμεγέθη ποιεῖ · στρυφνὸν δὲ τὸν μεγαλόσχημον, τραχὺν[1] τε καὶ πολυγώνιον καὶ ἀπεριφερῆ · ὀξὺν δὲ κατὰ τοὔνομα τὸν ὀξὺν τῷ ὄγκῳ καὶ κωνοειδῆ (γωνιοειδῆ Schneider) καὶ καμπύλον καὶ λεπτὸν καὶ ἀπεριφερῆ · δριμὺν δὲ τὸν περιφερῆ καὶ λεπτὸν καὶ γωνιοειδῆ καὶ καμπύλον · ἁλμυρὸν δὲ τὸν γωνιοειδῆ καὶ εὐμεγέθη (?) καὶ σκολιὸν καὶ ἰσοσκελῆ · πικρὸν δὲ τὸν περιφερῆ καὶ λεῖον, ἔχοντα σκολιότητα μέγεθος δὲ μικρόν · λιπαρὸν δὲ τὸν λεπτὸν καὶ στρογγύλον καὶ μικρόν[2] ('De Causis Plantarum,' vi. 1, ed. Schneider, 1818).

In another treatise Theophrastus again expounds at great

[1] Gassendi, as quoted below, evidently read τραχὺν δὲ τὸν πολυγώνιον.

[2] Gassendi translates thus:—

'Rotundas videlicet esse congruaque mole, quae Dulcem faciant; magna figura, quae Acerbum; multangula minimeque orbiculari, quae Asperum; acuta, conica, incurva, non tenui, non rotunda quae Acutum; orbiculata, tenui, angulata, incurva quae Acrem; angulata, distorta, crurumque aequalium, quae Salsum; rotunda, levi, distorta, parva quae Amarum; tenui, rotunda, parva quae Pinguem' ('Epicuri Philosophia,' 1675, vol. i., p. 156).

length Democritus's explanation of differences of taste, as caused by atoms of different shape touching the tongue. We quote the following:—

τὸν δὲ πικρὸν ἐκ μικρῶν καὶ λείων καὶ περιφερῶν, ἀλλ' ἐπ' ἐνίων μὲν σκαληνῶν, διὸ οὐδὲ πολυκάμπτων· βούλεται δὲ σκαληνὰ λέγειν, ἅπερ παράλλαξιν[1] ἔχει πρὸς ἄλληλα καὶ συμπλοκὴν ('De Sensu et Sensilibus,' c. 66).

Cicero thus refers to the atoms of Democritus:—

'Asperis et levibus et hamatis uncinatisque corporibus concreta haec esse' ('Acad.,' ii. 121). 'Esse corpuscula, quaedam levia, alia aspera, rotunda alia, partim angulata et pyramidata, curvata quaedam et quasi adunca' ('De Natura Deorum,' i. 66).

We may compare Gassendi:—

'Non posse quidem mentem assequi illam tantam varietatem figurarum, quae adscribendae sunt atomis; cum sint rotundae, ovatae, lenticulares, planae, gibbae, oblongae, turbinatae, hamatae, laeves, asperae, hispidae, tetrahedricae, pentahedricae, hexahedricae, &c., tam regulares quam irregulares, absque determinatione ulla intellectui possibili, ac potissimum irregularitatis formas commiscendo' (vol. i., p. 113).

Gassendi refutes the objection that atoms with slender projecting points must be liable to fracture, by saying that, since every part of the atom is perfectly solid and contains no void, atoms furnished with angles or apices or hooks are as indestructible as spherical atoms are.[2]

[1] Does παράλλαξιν πρὸς ἄλληλα mean 'overlapping each other'?

[2] 'Perspicuum est Atomos, qualescumque sint, corpuscula esse solidissima, inanisque plane expertia, quare et cum hami angulique eiusdem sint soliditatis, ... necesse est ipsos tam resistere ictibus externis, quam ipsa corpuscula orbicularia' (vol. i., p. 115).

IV. *Lucretius's argument for Free-will,* ii. 284-7.—*Is there a difficulty here?—'Necessity' in Nature and Freedom in man.*

> Quare in seminibus quoque idem fateare necessest,
> esse aliam praeter plagas et pondera causam
> motibus, unde haec est nobis innata potestas,
> de nilo quoniam fieri nil posse videmus. ii. 284-7.

'WHEREFORE in the case of atoms, too,[1] you must admit the same, namely, that besides blows and weights there is another cause for (their) movements,[2] whence this power of free action has been begotten in us, since we see that nothing can come from nothing.'

In these lines Lucretius sums up his reasoning on the most characteristic and weighty point of his master's system. The whole passage (ll. 251-93) is most closely reasoned. Not a word is thrown away.—When we come to this sentence, however, we pause and for a time are bewildered. Has not Lucretius told us that the atoms have two motions, a perpendicular downward motion, and a slight swerving from the perpendicular but for which they would never have come into contact? This swerving produces collisions among the atoms, or 'blows'—plagae. He has proved that blows could not have been but for declination. How then does he say that in

[1] The force of quoque must not be forgotten. It refers to the preceding illustrations of free-will action in men and animals. It means 'in atoms as well as in human beings.'

[2] Is motibus 'for their movements,' *i.e.*, for the movements of the atoms, or 'for our movements'? 'For their movements,' though less plausible, is the most consistent with Lucretius's argument. He is reasoning from men to the atoms, and applies his famous axiom ex nihilo nihil in a very bold and forcible way:—'If men can move at will, then the atoms which they come from must be able to move at will too.'

addition to weight and blows, which latter can only be caused by declination, we must admit the existence of declination? Is Lucretius unmindful or inconsistent? For does not this passage imply that plagae exist apart from declination and before it comes into play? Not by any means, we think. But the commentators certainly do not assist us to master the thought of this passage. In the first place, Cicero (referring possibly, his language leads us to think, to this very passage, which he may have read, and, if so, certainly misunderstood) has contrived to paraphrase the subject-matter of it in such a way as peculiarly to mislead any one who compares this passage of Lucretius with Cicero's words in 'De Fato.' He says, 'Epicurus declinatione atomi vitari fati necessitatem putat: itaque tertius quidam motus oritur extra pondus et plagam quum declinat atomus intervallo minimo' ('De Fato,' x.). Cicero here states the doctrine of Epicureanism in a singularly careless and inexact way, and his unqualified mode of applying the phrase, 'a third kind of motion,' seems to have misled all later commentators. In his note on the passage Mr. Munro makes no reference to the difficulty, but in his abstract of ll. 251-93 he gives the argument thus—'While the weight then of atoms enables *them* sometimes to withstand the external force of blows, it is only this declination of atoms at quite uncertain times and places which gives the mind its freedom of action,' and again on l. 288, 'Lucretius too, like Cicero, assigns the freedom of the will as the chief proof of the necessity of this third motion.' Again, M. Guyau ('La Morale d'Épicure,' p. 77), commenting on the passage, says, 'There exist then, according to Epicurus (and the testimony of Cicero here confirms that of Lucretius), three causes of motion, each profounder and more inward than the other; blows which are at the same time exterior and fatal (*fatal*), weight which is interior but appears still fatal, and finally Free-will which is at the same time interior and free.' And (p. 78, note), 'Cicero [in the passage above quoted] is entirely in agreement with Lucretius.'

Lucretius's reasoning becomes at once clear when we see that in this passage he is speaking *only* with reference to the human soul. He here assumes the existence of the world, as originally caused by Declination, and discusses the freedom of the will as a question entirely apart. He passes suddenly from the outer world governed by necessity in the form of natural laws (the consequence of pondus and plagae) to the soul of man. Lucretius is here insisting on *the freedom of the human will amid the vast mechanism of nature which surrounds it.* Man could not be free unless there exist in all atoms, and therefore in the atoms of his soul also, a principle apart from the pondus and plagae which govern the world without. This power of the soul-atoms to decline at will exists also in all atoms, but in the inorganic world he conceives it to be nullified. In the world of nature Epicurus knows of only two[1] causes of motion: first, Gravity causing a perpendicular, and secondly, Declination causing a swerving motion which produces plagae or 'collisions.' (It is plagae alone which, though there is no authority for so denoting it, might deserve to be called a third motion.) Strictly speaking, Cicero's phrase, 'a third kind of motion' (as applied by him and followed by all subsequent writers), is misleading. Free-will exists in all atoms. In the soul-atoms it is active and can originate motion, but in the atoms composing dead matter it is potential only, and can never be 'a cause of motion.' As we have already pointed out,[2] Epicurus seems to have assumed that the power of Declination, though still existing in the atoms, practically disappears after these atoms have combined to form matter.

[1] So far as we know, no ancient authority speaks of a third. Cf. Plutarch, 'De Plac. Ph.,' i. 23, 4,—Ἐπίκουρος δύο εἴδη τῆς κινήσεως, τὸ κατὰ στάθμην καὶ τὸ κατὰ παρέγκλισιν. Plutarch repeats this at i. 12.

[2] See p. 230.

Various counteracting causes tend to nullify it. Besides, Free-will is proportionally a far feebler power in gross matter than it is in the soul, and is far more easily held in check.

Epicurus speaks of no third cause of motion in the outer world. *It is only for the mind,* amid the necessity of nature which is twofold, *that a third cause of motion exists,* namely, the Free-will of the soul-atoms. This important paragraph can be understood aright only when we realize that in it Lucretius sharply distinguishes between the world of nature which is absolutely governed by necessity, that is to say, by natural law, and the mind of man.

We have always found a difficulty in this passage, as ordinarily explained, which we cannot think entirely of our own creation.

V. *Lucretius's reasoning in* ii. 251-93 *has been generally misunderstood. Does he conceive the laws of Nature as opposed to the 'foedera Fati'?—Mr. Benn's assertion that Lucretius grasps the fact of Law in Nature only from its negative side.*

IT is not easy to grasp Lucretius's reasoning throughout this paragraph, and not a few writers appear to have misunderstood it. In his admirable chapter on 'The Philosophy of Lucretius,' Professor Sellar observes that, according to Lucretius, creation is the result not of any Divine working, 'but of certain processes extending through infinite time, by means of which the atoms have at length been able to combine and work together in accordance with their ultimate conditions. The conception of these ultimate conditions and of their relations to one another involves some more vital agency than that of blind chance or an iron fatalism (ii. 254). The foedera naturai are opposed to the foedera fati. The idea of law in Nature, as understood by Lucretius, is not merely that of invariable sequence or concomitance of phenomena. It implies at least the further idea of a "secreta facultas"[1] in the original elements' (p. 335). The most careful study of all the doctrines of Lucretius's system and their bearing on each other shows us no ground for admitting any opposition between foedera fati and foedera naturai. Lucretius, it is true, does not believe in Fate, so far as men are concerned. In the moral world he asserts that there is no such thing. At the same time, 'Fate' or 'Necessity' is a name occasionally given, as we have seen,[2] both by Epicurus and Lucretius to the order of Nature resulting from natural laws. The foedera fati (a mere synonym for fatum) and the foedera naturai are never really opposed to each other by Lucretius. Such a conception is altogether foreign to him.

Again, Mr. Alfred Benn, in an able article on 'Epicurus and Lucretius' in the 'Westminster Review' (April, 1882), insists

[1] At p. 319, Professor Sellar says: 'A secret faculty in the atoms, distinct from their other properties, is assumed. Thus he says—

> At primordia gignundis in rebus oportet
> naturam clandestinam coecamque adhibere. i. 778-9.'

This quotation is translated as follows in the note: 'But it is necessary that the atoms, in the act of creation, should exercise some secret, invisible faculty.' Putting aside the fact that *secreta facultas* (a phrase occurring only once in the poem, at i. 178) cannot possibly mean a 'secret faculty,' and that i. 778-9 means, as Mr. Munro has shown, merely that the atoms must not possess any secondary qualities such as colour, the expressions used by Professor Sellar are not consistent with Lucretius's system. His atoms possess no properties apart from those which he assigns them: figure, perfect-hardness, &c., and also Free-will. How then can we find room within the rigid four walls of Epicureanism for anything like a 'vital agency,' either as working in Nature or as finding expression in the laws of Nature? Instead of this, how often does Lucretius tell us that the origin and the maintenance of the world and its life is due to a mere coincidence among the atoms?

[2] See page 221, notes 1 and 2, where all the references to fatum are collated. Cf. especially v. 309-10, where fati finis, 'the limits of fate,' evidently refers to the same thing as naturae foedera.

repeatedly that Epicurus has no title to the credit of asserting the reign of Law. He says that the Stoics have more claim to this honour, and in their physics 'came nearer than Lucretius to the standpoint of modern science,' and even asserts that 'Epicurus expressly refused to accept such a doctrine' (the universality of law in nature). Mr. Benn brings little evidence to support this remarkable statement. Probably it is based in part on a misconception of Epicurus's doctrine of Atomic Declination. Referring to the latter, he says, 'Apparently neither Epicurus nor his disciples saw that in discarding the invariable sequence of phenomena, they annulled to the same extent the possibility of human foresight and the adaptation of means to ends' (p. 323). The writer, possibly under the influence of M. Guyau, assumes that the consequence of Free-will existing in the atoms must be a power of spontaneous movement in all material substances, which must interfere with the regular order of nature. But, as we saw, Epicurus held that Free-will, though active in the atoms, is nullified when these combine in matter. Thus it did not, according to Epicurus's conception of it, at all interfere with Law. Again he says, 'Lucretius expressly tells us (ii. 255, ex infinito ne causam causa sequatur) that the law of causation is broken through by the clinamen.' The writer here fails to see that Lucretius draws a sharp distinction between the world of nature, subject to law, and the human mind, which is free. So far as nature (that is, the method of the world's ongoings) is concerned, without taking into account the agency of man, Lucretius holds that causam causa sequitur, 'cause *does* follow cause.' The truth is that Lucretius had the firmest grasp of the fact of Law. At the same time he holds that the mind of man is not subject to the foedera naturae. Free-will is a libera potestas. But perhaps Mr. Benn holds that a belief in Free-will is not consistent with a belief in Laws of Nature. This would help us to understand his assertion that Epicurus did not to any extent believe in

Law. Again he says (p. 333), that 'when Lucretius speaks of foedera naturae, he means not what we understand by laws of nature' . . . 'but rather the limiting possibilities of existence.' In fact, Mr. Benn holds that Lucretius grasped merely the negative side of natural order. A less fair criticism than this could hardly be made. The majestas cognita rerum which so inspired Lucretius was something more than 'negative' knowledge.

CHISWICK PRESS :—C. WHITTINGHAM AND CO., TOOKS COURT, CHANCERY LANE.

August 1888.

A CLASSIFIED LIST

OF

EDUCATIONAL WORKS

PUBLISHED BY

GEORGE BELL & SONS.

Full Catalogues will be sent post free on application.

BIBLIOTHECA CLASSICA.

A Series of Greek and Latin Authors, with English Notes, edited by eminent Scholars. 8vo.

Æschylus. By F. A. Paley, M.A., LL.D. 18*s*.
Cicero's Orations. By G. Long, M.A. 4 vols. 16*s*., 14*s*., 16*s*., 18*s*.
Demosthenes. By R. Whiston, M.A. 2 vols. 16*s*. each.
Euripides. By F. A. Paley, M.A., LL.D. 3 vols. 16*s*. each.
Homer. By F. A. Paley, M.A., LL.D. Vol. I. 12*s*.; Vol. II. 14*s*.
Herodotus. By Rev. J. W. Blakesley, B.D. 2 vols. 32*s*.
Hesiod. By F. A. Paley, M.A., LL.D. 10*s*. 6*d*.
Horace. By Rev. A. J. Macleane, M.A. 18*s*.
Juvenal and Persius. By Rev. A. J. Macleane, M.A. 12*s*.
Lucan. The Pharsalia. By C. E. Haskins, M.A. 14*s*.
Plato. By W. H. Thompson, D.D. 2 vols. 7*s*. 6*d*. each.
Sophocles. Vol. I. By Rev. F. H. Blaydes, M.A. 18*s*.
——— Vol. II. Philoctetes—Electra—Ajax and Trachiniæ. By F. A. Paley, M.A., LL.D. 12*s*. Or the 4 plays separately, 3*s*. 6*d*. each.
Tacitus: The Annals. By the Rev. P. Frost. 15*s*.
Terence. By E. St. J. Parry, M.A. 18*s*.
Virgil. By J. Conington, M.A. Revised by Professor H. Nettleship. 3 vols. 14*s*. each.
An Atlas of Classical Geography; Twenty-four Maps. By W. Hughes and George Long, M.A. New edition, with coloured Outlines. Imperial 8vo. 12*s*. 6*d*.

Uniform with above.

A Complete Latin Grammar. By J. W. Donaldson, D.D. 3rd Edition. 14*s*.

GRAMMAR-SCHOOL CLASSICS.

A Series of Greek and Latin Authors, with English Notes.
Fcap. 8vo.

Cæsar: De Bello Gallico. By George Long, M.A. 4*s*.

——— Books I.–III. For Junior Classes. By G. Long, M.A. 1*s*. 6*d*.

——— Books IV. and V. 1*s*. 6*d*. Books VI. and VII. 1*s*. 6*d*.

Catullus, Tibullus, and Propertius. Selected Poems. With Life. By Rev. A. H. Wratislaw. 2*s*. 6*d*.

Cicero: De Senectute, De Amicitia, and Select Epistles. By George Long, M.A. 3*s*.

Cornelius Nepos. By Rev. J. F. Macmichael. 2*s*.

Homer: Iliad. Books I.–XII. By F. A. Paley, M.A., LL.D. 4*s*. 6*d*. Also in 2 parts, 2*s*. 6*d*. each.

Horace. With Life. By A. J. Macleane, M.A. 3*s*. 6*d*. In 2 parts, 2*s*. each.

Juvenal: Sixteen Satires. By H. Prior, M.A. 3*s*. 6*d*.

Martial: Select Epigrams. With Life. By F. A. Paley, M.A., LL.D. 4*s*. 6*d*.

Ovid: the Fasti. By F. A. Paley, M.A., LL.D. 3*s*. 6*d*. Books I. and II. 1*s*. 6*d*. Books III. and IV. 1*s*. 6*d*.

Sallust: Catilina and Jugurtha. With Life. By G. Long, M.A. and J. G. Frazer. 3*s*. 6*d*., or separately, 2*s*. each.

Tacitus: Germania and Agricola. By Rev. P. Frost. 2*s*. 6*d*.

Virgil: Bucolics, Georgics, and Æneid, Books I.–IV. Abridged from Professor Conington's Edition. 4*s*. 6*d*.—Æneid, Books V.–XII. 4*s*. 6*d*. Also in 9 separate Volumes, as follows, 1*s*. 6*d*. each:—Bucolics—Georgics, I. and II.—Georgics, III. and IV.—Æneid, I. and II.—Æneid, III. and IV.—Æneid, V. and VI.—Æneid, VII. and VIII.—Æneid, IX. and X.—Æneid, XI. and XII.

Xenophon: The Anabasis. With Life. By Rev. J. F. Macmichael. 3*s*. 6*d*. Also in 4 separate volumes, 1*s*. 6*d*. each:—Book I. (with Life, Introduction, Itinerary, and Three Maps)—Books II. and III.—IV. and V.—VI. and VII.

——— The Cyropædia. By G. M. Gorham, M.A. 3*s*. 6*d*. Books I. and II. 1*s*. 6*d*.—Books V. and VI. 1*s*. 6*d*.

——— Memorabilia. By Percival Frost, M.A. 3*s*.

A Grammar-School Atlas of Classical Geography, containing Ten selected Maps. Imperial 8vo. 5*s*.

Uniform with the Series.

The New Testament, in Greek. With English Notes, &c. By Rev. J. F. Macmichael. 4*s*. 6*d*. In parts, sewed, 6*d*. each.

CAMBRIDGE GREEK AND LATIN TEXTS.

Æschylus. By F. A. Paley, M.A., LL.D. 2*s*. 6*d*.

Cæsar: **De Bello Gallico.** By G. Long, M.A. 1*s*. 6*d*.

Cicero: De Senectute et De Amicitia, et Epistolæ Selectæ. By G. Long, M.A. 1*s*. 6*d*.

Ciceronis Orationes. In Verrem. By G. Long, M.A. 2*s*. 6*d*.

Euripides. By F. A. Paley, M.A., LL.D. 3 vols. 2*s*. each.

Herodotus. By J. G. Blakesley, B.D. 2 vols. 5*s*.

Homeri Ilias. I.–XII. By F. A. Paley, M.A., LL.D. 1*s*. 6*d*.

Horatius. By A. J. Macleane, M.A. 1*s*. 6*d*.

Juvenal et Persius. By A. J. Macleane, M.A. 1*s*. 6*d*.

Lucretius. By H. A. J. Munro, M.A. 2*s*.

Sallusti Crispi Catilina et Jugurtha. By G. Long, M.A. 1*s*. 6*d*.

Sophocles. By F. A. Paley, M.A., LL.D. 2*s*. 6*d*.

Terenti Comœdiæ. By W. Wagner, Ph.D. 2*s*.

Thucydides. By J. G. Donaldson, D.D. 2 vols. 4*s*.

Virgilius. By J. Conington, M.A. 2*s*.

Xenophontis Expeditio Cyri. By J. F. Macmichael, B.A. 1*s*. 6*d*.

Novum Testamentum Græce. By F. H. Scrivener, M.A., D.C.L. 4*s*. 6*d*. An edition with wide margin for notes, half bound, 12*s*. EDITIO MAJOR, with additional Readings and References. 7*s*. 6*d*. *See page* 9.

CAMBRIDGE TEXTS WITH NOTES.

A Selection of the most usually read of the Greek and Latin Authors, Annotated for Schools. Edited by well-known Classical Scholars. Fcap. 8vo. 1*s*. 6*d*. *each, with exceptions.*

Æschylus. Prometheus Vinctus.—Septem contra Thebas.—Agamemnon.—Persæ.—Eumenides.—Choephoroe. By F. A. Paley, M.A., LL.D.

Euripides. Alcestis.—Medea.—Hippolytus.—Hecuba.—Bacchæ.—Ion. 2*s*.—Orestes.—Phoenissæ.—Troades.—Hercules Furens.—Andromache.—Iphigenia in Tauris.—Supplices. By F. A. Paley, M.A., LL.D.

Homer. Iliad. Book I. By F. A. Paley, M.A., LL.D. 1*s*.

Sophocles. Œdipus Tyrannus.—Œdipus Coloneus.—Antigone.—Electra—Ajax. By F. A. Paley, M.A., LL.D.

Xenophon. Anabasis. In 4 vols. By J. E. Melhuish, M.A., Assistant Classical Master at St. Paul's School.

Xenophon. Hellenics, Book II. By L. D. Dowdall, M.A., B.D. [*Shortly.*

Cicero. De Senectute, De Amicitia, and Epistolæ Selectæ. By G. Long, M.A.

Ovid. Fasti. By F. A. Paley, M.A., LL.D. In 3 vols., 2 books in each. 2*s*. each vol.

Ovid. Selections. Amores, Tristia, Heroides, Metamorphoses. By A. J. Macleane, M.A.

Terence. Andria.—Hauton Timorumenos.—Phormio.—Adelphoe. By Professor Wagner, Ph.D.

Virgil. Professor Conington's edition, abridged in 9 vols., 2 books in each.

Others in preparation.

PUBLIC SCHOOL SERIES.

A Series of Classical Texts, annotated by well-known Scholars. Cr. 8vo.

Aristophanes. The Peace. By F. A. Paley, M.A., LL.D. 4*s*. 6*d*.
——— The Acharnians. By F. A. Paley, M.A., LL.D. 4*s*. 6*d*.
——— The Frogs. By F. A. Paley, M.A., LL.D. 4*s*. 6*d*.
Cicero. The Letters to Atticus. Bk. I. By A. Pretor, M.A. 4*s*. 6*d*.
Demosthenes de Falsa Legatione. By R. Shilleto, M.A. 6*s*.
——— The Law of Leptines. By B. W. Beatson, M.A. 3*s*. 6*d*.
Livy. Book XXI. Edited, with Introduction, Notes, and Maps, by the Rev. L. D. Dowdall, M.A., B.D. 3s. 6d. [*Book XXII. in the press.*
——— Book XXII. Edited, &c., by Rev. L. D. Dowdall, M.A., B.D. 3s. 6d.
Plato. The Apology of Socrates and Crito. By W. Wagner, Ph.D. 10th Edition. 3s. 6d. Cheap Edition, limp cloth, 2s. 6d.
——— The Phædo. 9th Edition. By W. Wagner, Ph.D. 5*s*. 6*d*.
——— The Protagoras. 4th Edition. By W. Wayte, M.A. 4*s*. 6*d*.
——— The Euthyphro. 3rd Edition. By G. H. Wells, M.A. 3*s*.
——— The Euthydemus. By G. H. Wells, M.A. 4*s*.
——— The Republic. Books I. & II. By G. H. Wells, M.A. 3rd Edition. 5s. 6d.
Plautus. The Aulularia. By W. Wagner, Ph.D. 3rd Edition. 4*s*. 6*d*.
——— Trinummus. By W. Wagner, Ph.D. 3rd Edition. 4*s*. 6*d*.
——— The Menaechmei. By W. Wagner, Ph.D. 2nd Edit. 4*s*. 6*d*.
——— The Mostellaria. By Prof. E. A. Sonnenschein. 5*s*.
——— The Rudens. Edited by Prof. E. A. Sonnenschein. [*In the press.*
Sophoclis Trachiniæ. By A. Pretor, M.A. 4*s*. 6*d*.
Sophocles. Oedipus Tyrannus. By B. H. Kennedy, D.D. 5*s*.
Terence. By W. Wagner, Ph.D. 2nd Edition. 10*s*. 6*d*.
Theocritus. By F. A. Paley, M.A., LL.D. 2nd Edition. 4*s*. 6*d*.
Thucydides. Book VI. By T. W. Dougan, M.A., Fellow of St. John's College, Cambridge. 6s.

Others in preparation.

CRITICAL AND ANNOTATED EDITIONS.

Aristophanis Comœdiæ. By H. A. Holden, LL.D. 8vo. 2 vols. 23s. 6d. Plays sold separately.

Calpurnius Siculus. By C. H. Keene, M.A. Crown 8vo. 6*s*.

Corpus Poetarum Latinorum. Edited by Walker. 1 vol. 8vo. 18*s*.

Horace. Quinti Horatii Flacci Opera. By H. A. J. Munro, M.A. Large 8vo. 1*l.* 1*s.*

Livy. The first five Books. By J. Prendeville. 12mo. roan, 5*s.* Or Books I.-III. 3*s.* 6*d.* IV. and V. 3*s.* 6*d.*

Lucretius. With Commentary by H. A. J. Munro. 4th Edition. Vols. I. and II. Introduction, Text, and Notes. 18*s.* Vol. III. Translation. 6*s.*

Ovid. P. Ovidii Nasonis Heroides XIV. By A. Palmer, M.A. 8vo. 6*s.*

——— P. Ovidii Nasonis Ars Amatoria et Amores. By the Rev. H. Williams, M.A. 3*s.* 6*d.*

——— Metamorphoses. Book XIII. By Chas. Haines Keene, M.A. 2*s.* 6*d.*

——— Epistolarum ex Ponto Liber Primus. By C.H. Keene, M.A. 3*s.*

Propertius. Sex Aurelii Propertii Carmina. By F. A. Paley, M.A., LL.D. 8vo. Cloth, 9*s.*

——— Sex Propertii Elegiarum. Libri IV. Recensuit A. Palmer, Collegii Sacrosanctæ et Individuæ Trinitatis juxta Dublinum Socius. Fcap 8vo. 3*s.* 6*d.*

Sophocles. The Ajax. By C. E. Palmer, M.A. 3*s.* 6*d.*

——— The Oedipus Tyrannus. By B. H. Kennedy, D.D. Crown 8vo. 8*s.*

Thucydides. The History of the Peloponnesian War. By Richard Shilleto, M.A. Book I. 8vo. 6*s.* 6*d.* Book II. 8vo. 5*s.* 6*d.*

LATIN AND GREEK CLASS-BOOKS.

Faciliora. An Elementary Latin Book on a new principle. By the Rev. J. L. Seager, M.A. [*In the press.*

First Latin Lessons. By A. M. M. Stedman, M.A. 1*s.*

Miscellaneous Latin Exercises. By A. M. M. Stedman, M.A. Fcap. 8vo. 1*s.* 6*d.*

Easy Latin Passages for Unseen Translation. By A. M. M. Stedman, M.A. Fcap. 8vo. 1*s.* 6*d.*

A Latin Primer. By Rev. A. C. Clapin, M.A. 1*s.*

Auxilia Latina. A Series of Progressive Latin Exercises. By M. J. B. Baddeley, M.A. Fcap. 8vo. Part I. Accidence. 3rd Edition, revised. 2*s.* Part II. 4th Edition, revised. 2*s.* Key to Part II. 2*s.* 6*d.*

Scala Latina. Elementary Latin Exercises. By Rev. J. W. Davis, M.A. New Edition, with Vocabulary. Fcap. 8vo. 2*s.* 6*d.*

Latin Prose Lessons. By Prof. Church, M.A. 8th Edition. Fcap. 8vo. 2*s.* 6*d.*

Latin Exercises and Grammar Papers. By T. Collins, M.A. 5th Edition. Fcap. 8vo. 2*s.* 6*d.*

Unseen Papers in Latin Prose and Verse. With Examination Questions. By T. Collins, M.A. 4th Edition. Fcap. 8vo. 2*s.* 6*d.*

——— in Greek Prose and Verse. With Examination Questions. By T. Collins, M.A. 3rd Edition. Fcap. 8vo. 3*s.*

Easy Translations from Nepos, Cæsar, Cicero, Livy, &c., for Retranslation into Latin. With Notes, by T. Collins, M.A. [*Preparing.*

Tales for Latin Prose Composition. With Notes and Vocabulary. By G. H. Wells, M.A. 2*s.*

Latin Vocabularies for Repetition. By A. M. M. Stedman, M.A. 2nd Edition, revised. Fcap. 8vo. 1s. 6*d*.

Analytical Latin Exercises. By C. P. Mason, B.A. 4th Edit. Part I., 1s. 6*d*. Part II., 2s. 6*d*.

Latin Examination Papers in Grammar and Idiom. By A. M. M. Stedman, M.A. Crown 8vo. 2s. 6*d*. Key (for Tutors and Private Students only), 5s.

Greek Examination Papers in Grammar and Idiom. By A. M. M. Stedman, M.A. 2s. 6*d*.

Scala Græca: a Series of Elementary Greek Exercises. By Rev. J. W. Davis, M.A., and R. W. Baddeley, M.A. 3rd Edition. Fcap. 8vo. 2s. 6*d*.

Greek Verse Composition. By G. Preston, M.A. 5th Edition. Crown 8vo. 4s. 6*d*.

Greek Particles and their Combinations according to Attic Usage. A Short Treatise. By F. A. Paley, M.A., LL.D. 2s. 6*d*.

Greek Testament Selections. By A. M. M. Stedman, M.A. 2nd Edition enlarged, with Notes and Vocabulary. Fcap. 8vo. 2s. 6*d*.

Rudiments of Attic Construction and Idiom. By the Rev. W. C. Compton, M.A., Assistant Master at Uppingham School. 3s.

Passages for Translation into Latin Prose. By Prof. H. Nettleship, M.A. 3s. Key (for Tutors only), 4s. 6*d*.

BY THE REV. P. FROST, M.A., ST. JOHN'S COLLEGE, CAMBRIDGE.

Eclogæ Latinæ; or, First Latin Reading-Book, with English Notes and a Dictionary. New Edition. Fcap. 8vo. 1s. 6*d*.

Materials for Latin Prose Composition. New Edition. Fcap. 8vo. 2s. Key (for Tutors only), 4s.

A Latin Verse-Book. An Introductory Work on Hexameters and Pentameters. New Edition. Fcap. 8vo. 2s. Key (for Tutors only), 5s.

Analecta Græca Minora, with Introductory Sentences, English Notes, and a Dictionary. New Edition. Fcap. 8vo. 2s.

Materials for Greek Prose Composition. New Edit. Fcap. 8vo. 2s. 6*d*. Key (for Tutors only), 5s.

Florilegium Poeticum. Elegiac Extracts from Ovid and Tibullus. New Edition. With Notes. Fcap. 8vo. 2s.

Anthologia Græca. A Selection of Choice Greek Poetry, with Notes. By F. St. John Thackeray. *4th and Cheaper Edition.* 16mo. 4s. 6*d*.

Anthologia Latina. A Selection of Choice Latin Poetry, from Nævius to Boëthius, with Notes. By Rev. F. St. John Thackeray. Revised and Cheaper Edition. 16mo. 4s. 6*d*.

BY H. A. HOLDEN, LL.D.

Foliorum Silvula. Part I. Passages for Translation into Latin Elegiac and Heroic Verse. 10th Edition. Post 8vo. 7s. 6*d*.

——— Part II. Select Passages for Translation into Latin Lyric and Comic Iambic Verse. 3rd Edition. Post 8vo. 5s.

Folia Silvulæ, sive Eclogæ Poetarum Anglicorum in Latinum et Græcum conversæ. 8vo. Vol. II. 12s.

Foliorum Centuriæ. Select Passages for Translation into Latin and Greek Prose. 10th Edition. Post 8vo. 8s.

TRANSLATIONS, SELECTIONS, &c.

*** Many of the following books are well adapted for School Prizes.

Æschylus. Translated into English Prose by F. A. Paley, M.A., LL.D. 2nd Edition. 8vo. 7*s*. 6*d*.

——— Translated into English Verse by Anna Swanwick. 4th Edition. Post 8vo. 5*s*.

Horace. The Odes and Carmen Sæculare. In English Verse by J. Conington, M.A. 10th edition. Fcap. 8vo. 5*s*. 6*d*.

——— The Satires and Epistles. In English Verse by J. Conington, M.A. 7th edition. 6*s*. 6*d*.

Plato. Gorgias. Translated by E. M. Cope, M.A. 8vo. 2nd Ed. 7*s*.

——— Philebus. Trans. by F. A. Paley, M.A., LL.D. Sm. 8vo. 4*s*.

——— Theætetus. Trans. by F. A. Paley, M.A., LL.D. Sm. 8vo. 4*s*.

——— Analysis and Index of the Dialogues. By Dr. Day. Post 8vo. 5*s*.

Sophocles. Œdipus Tyrannus. By Dr. Kennedy. 1*s*.

Theocritus. In English Verse, by C. S. Calverley, M.A. New Edition, revised. Crown 8vo. 7*s*. 6*d*.

Translations into English and Latin. By C. S. Calverley, M.A. Post 8vo. 7*s*. 6*d*.

Translations into English, Latin, and Greek. By R. C. Jebb, M.A., H. Jackson, Litt.D., and W. E. Currey, M.A. Second Edition. 8*s*.

Extracts for Translation. By R. C. Jebb, M.A., H. Jackson, Litt.D., and W. E. Currey, M.A. 4*s*. 6*d*.

Between Whiles. Translations by Rev. B. H. Kennedy, D.D. 2nd Edition, revised. Crown 8vo. 5*s*.

REFERENCE VOLUMES.

A Latin Grammar. By Albert Harkness. Post 8vo. 6*s*.

——— By T. H. Key, M.A. 6th Thousand. Post 8vo. 8*s*.

A Short Latin Grammar for Schools. By T. H. Key, M.A. F.R.S. 16th Edition. Post 8vo. 3*s*. 6*d*.

A Guide to the Choice of Classical Books. By J. B. Mayor, M.A. 3rd Edition, with a Supplementary List. Crown 8vo. 4*s*. 6*d*. Supplementary List separately, 1*s*. 6*d*.

The Theatre of the Greeks. By J. W. Donaldson, D.D. 8th Edition. Post 8vo. 5*s*.

Keightley's Mythology of Greece and Italy. 4th Edition. 5*s*.

CLASSICAL TABLES.

Latin Accidence. By the Rev. P. Frost, M.A. 1*s.*

Latin Versification. 1*s.*

Notabilia Quædam; or the Principal Tenses of most of the Irregular Greek Verbs and Elementary Greek, Latin, and French Construction. New Edition. 1s.

Richmond Rules for the Ovidian Distich, &c. By J. Tate, M.A. 1s.

The Principles of Latin Syntax. 1*s.*

Greek Verbs. A Catalogue of Verbs, Irregular and Defective; their leading formations, tenses, and inflexions, with Paradigms for conjugation, Rules for formation of tenses, &c. &c. By J. S. Baird, T.C.D. 8th Ed. 2s. 6d.

Greek Accents (Notes on). By A. Barry, D.D. New Edition. 1*s.*

Homeric Dialect. Its Leading Forms and Peculiarities. By J. S. Baird, T.C.D. New Edition, by W. G. Rutherford, LL.D. 1s.

Greek Accidence. By the Rev. P. Frost, M.A. New Edition. 1*s.*

CAMBRIDGE MATHEMATICAL SERIES.

Arithmetic for Schools. By C. Pendlebury, M.A. 2nd Edition, revised, 4s. 6d. Or in two parts, 2s. 6d. each. Examples (nearly 8000), without answers, in a separate vol. 3s.

Algebra. Choice and Chance. By W. A. Whitworth, M.A. 4th Edition. 6s.

Euclid. Books I.–VI. and part of Books XI. and XII. By H. Deighton. 4s. 6d. Books I. and II., 2s. [*A Key to the Exercises in the press.*

Euclid. Exercises on Euclid and in Modern Geometry. By J. McDowell, M.A. 3rd Edition. 6s.

Trigonometry. Plane. By Rev. T. Vyvyan, M.A. 3rd Edit. 3*s.* 6*d.*

Geometrical Conic Sections. By H. G. Willis, M.A. Manchester Grammar School. 7s. 6d.

Conics. The Elementary Geometry of. 5th Edition, revised and enlarged. By C. Taylor, D.D. 4s. 6d.

Solid Geometry. By W. S. Aldis, M.A. 4th Edit. revised. 6*s.*

Geometrical Optics. By W. S. Aldis, M.A. 2nd Edition. 4*s.*

Rigid Dynamics. By W. S. Aldis, M.A. 4*s.*

Elementary Dynamics. By W. Garnett, M.A., D.C.L. 4th Ed. 6*s.*

Dynamics. A Treatise on. By W. H. Besant, D.Sc., F.R.S. 7*s.* 6*d.*

Heat. An Elementary Treatise. By W. Garnett, M.A., D.C.L. 4th Edition. 4s.

Elementary Physics. Examples in. By W. Gallatly, M.A. [*In the press.*

Hydromechanics. By W. H. Besant, D.Sc., F.R.S. 4th Edition. Part I. Hydrostatics. 5s.

Mathematical Examples. By J. M. Dyer, M.A., and R. Prowde Smith, M.A., Assistant Masters at Cheltenham College. [*In the press.*

Mechanics. Problems in Elementary. By W. Walton, M.A. 6*s.*

CAMBRIDGE SCHOOL AND COLLEGE TEXT-BOOKS.

A Series of Elementary Treatises for the use of Students.

Arithmetic. By Rev. C. Elsee, M.A. Fcap. 8vo. 13th Edit. 3*s*. 6*d*.

——— By A. Wrigley, M.A. 3*s*. 6*d*.

——— A Progressive Course of Examples. With Answers. By J. Watson, M.A. 6th Edition. 2*s*. 6*d*.

Algebra. By the Rev. C. Elsee, M.A. 7th Edit. 4*s*.

——— Progressive Course of Examples. By Rev. W. F. M'Michael, M.A., and R. Prowde Smith, M.A. 4th Edition. 3*s*. 6*d*. With Answers. 4*s*. 6*d*.

Plane Astronomy, An Introduction to. By P. T. Main, M.A 5th Edition. 4*s*.

Conic Sections treated Geometrically. By W. H. Besant, D.Sc. 6th Edition. 4*s*. 6*d*. Solution to the Examples. 4*s*.

——— Enunciations and Figures Separately. 1*s*. 6*d*.

Statics, Elementary. By Rev. H. Goodwin, D.D. 2nd Edit. 3*s*.

Hydrostatics, Elementary. By W. H. Besant, D.Sc. 12th Edit. 4*s*.

Mensuration, An Elementary Treatise on. By B. T. Moore, M.A. 5*s*.

Newton's Principia, The First Three Sections of, with an Appendix; and the Ninth and Eleventh Sections. By J. H. Evans, M.A. 5th Edition, by P. T. Main, M.A. 4*s*.

Analytical Geometry for Schools. By T. G. Vyvyan. 4th Edit. 4*s*. 6*d*.

Greek Testament, Companion to the. By A. C. Barrett, A.M. 5th Edition, revised. Fcap. 8vo. 5*s*.

Book of Common Prayer, An Historical and Explanatory Treatise on the. By W. G. Humphry, B.D. 6th Edition. Fcap. 8vo. 2*s*. 6*d*.

Music, Text-book of. By Professor H. C. Banister. 13th Edition, revised. 5*s*.

——— Concise History of. By Rev. H. G. Bonavia Hunt, Mus. Doc. Dublin. 9th Edition revised. 3*s*. 6*d*.

ARITHMETIC AND ALGEBRA.

See the two foregoing Series.

BOOK-KEEPING.

Book-keeping Papers, set at various Public Examinations. Collected and Written by J. T. Medhurst, Lecturer on Book-keeping in the City of London College.

GEOMETRY AND EUCLID.

Euclid. Books I.–VI. and part of XI. and XII. A New Translation. By H. Deighton. (See Cambridge Mathematical Series, p. 8.)

——— The Definitions of, with Explanations and Exercises, and an Appendix of Exercises on the First Book. By R. Webb, M.A. Crown 8vo. 1s. 6d.

——— Book I. With Notes and Exercises for the use of Preparatory Schools, &c. By Braithwaite Arnett, M.A. 8vo. 4s. 6d.

——— The First Two Books explained to Beginners. By C. P. Mason, B.A. 2nd Edition. Fcap. 8vo. 2s. 6d.

The Enunciations and Figures to Euclid's Elements. By Rev. J. Brasse, D.D. New Edition. Fcap. 8vo. 1s. Without the Figures, 6d.

Exercises on Euclid and in Modern Geometry. By J. McDowell, B.A. Crown 8vo. 3rd Edition revised. 6s.

Geometrical Conic Sections. By H. G. Willis, M.A. (See p. 8.)

Geometrical Conic Sections. By W. H. Besant, D.Sc. (See p. 9.)

Elementary Geometry of Conics. By C. Taylor, D.D. (See p. 8.)

An Introduction to Ancient and Modern Geometry of Conics. By C. Taylor, D.D., Master of St. John's Coll., Camb. 8vo. 15s.

Solutions of Geometrical Problems, proposed at St. John's College from 1830 to 1846. By T. Gaskin, M.A. 8vo. 12s.

TRIGONOMETRY.

Trigonometry, Introduction to Plane. By Rev. T. G. Vyvyan, Charterhouse. 3rd Edition. Cr. 8vo. 3s. 6d.

An Elementary Treatise on Mensuration. By B. T. Moore, M.A. 5s.

ANALYTICAL GEOMETRY AND DIFFERENTIAL CALCULUS.

An Introduction to Analytical Plane Geometry. By W. P. Turnbull, M.A. 8vo. 12s.

Problems on the Principles of Plane **Co-ordinate Geometry.** By W. Walton, M.A. 8vo. 16s.

Trilinear Co-ordinates, and Modern Analytical Geometry of Two Dimensions. By W. A. Whitworth, M.A. 8vo. 16s.

An Elementary Treatise on Solid Geometry. By W. S. Aldis, M.A. 4th Edition revised. Cr. 8vo. 6s.

Elliptic Functions, Elementary Treatise on. By A. Cayley, D.Sc. Professor of Pure Mathematics at Cambridge Universiity. Demy 8vo. 15s.

MECHANICS & NATURAL PHILOSOPHY.

Statics, Elementary. By H. Goodwin, D.D. Fcap. 8vo. 2nd Edition. 3s.

Dynamics, A Treatise on Elementary. By W. Garnett, M.A., D.C.L. 4th Edition. Crown 8vo. 6s.

Dynamics. Rigid. By W. S. Aldis, M.A. 4s.

Dynamics. A Treatise on. By W. H. Besant, D.Sc., F.R.S. 7s. 6d.

Elementary Mechanics, Problems in. By W. Walton, M.A. New Edition. Crown 8vo. 6s.

Theoretical Mechanics, Problems in. By W. Walton, M.A. 3rd Edition. Demy 8vo. 16s.

Hydrostatics. By W. H. Besant, D.Sc. Fcap. 8vo. 12th Edition. 4s.

Hydromechanics, A Treatise on. By W. H. Besant, D.Sc., F.R.S. 8vo. 4th Edition, revised. Part I. Hydrostatics. 5s.

Optics, Geometrical. By W. S. Aldis, M.A. Crown 8vo. 2nd Edition. 4s.

Double Refraction, A Chapter on Fresnel's Theory of. By W. S. Aldis, M.A. 8vo. 2s.

Heat, An Elementary Treatise on. By W. Garnett, M.A., D.C.L. Crown 8vo. 4th Edition. 4s.

Newton's Principia, The First Three Sections of, with an Appendix; and the Ninth and Eleventh Sections. By J. H. Evans, M.A. 5th Edition. Edited by P. T. Main, M.A. 4s.

Astronomy, An Introduction to Plane. By P. T. Main, M.A. Fcap. 8vo. cloth. 5th Edition. 4s.

——— Practical and Spherical. By R. Main, M.A. 8vo. 14s.,

——— Elementary Chapters on, from the 'Astronomie Physique' of Biot. By H. Goodwin, D.D. 8vo. 3s. 6d.

Pure Mathematics and Natural Philosophy, A Compendium of Facts and Formulæ in. By G. R. Smalley. 2nd Edition, revised by J. McDowell, M.A. Fcap. 8vo. 3s. 6d.

Elementary Mathematical Formulæ. By the Rev. T. W. Openshaw, M.A. 1s. 6d.

Elementary Course of Mathematics. By H. Goodwin, D.D. 6th Edition. 8vo. 16s.

Problems and Examples, adapted to the 'Elementary Course of Mathematics.' 3rd Edition. 8vo. 5s.

Solutions of Goodwin's Collection of Problems and Examples. By W. W. Hutt, M.A. 3rd Edition, revised and enlarged. 8vo. 9s.

A Collection of Examples and Problems in Arithmetic, Algebra, Geometry, Logarithms, Trigonometry, Conic Sections, Mechanics, &c., with Answers. By Rev. A. Wrigley. 20th Thousand. 8s. 6d. Key. 10s. 6d.

TECHNOLOGICAL HANDBOOKS.

Edited by H. Trueman Wood, Secretary of the Society of Arts.

1. **Dyeing and Tissue Printing.** By W. Crookes, F.R.S. 5s.
2. **Glass Manufacture.** By Henry Chance, M.A.; H. J. Powell, B.A.; and H. G. Harris. 3s. 6d.
3. **Cotton Manufacture.** By Richard Marsden, of Manchester. 3rd Edition, revised. 6s. 6d.
4. **Chemistry of Coal-Tar Colours.** By Prof. Benedikt. Translated by Dr. Knecht of Bradford. 5s.
5. **Woollen and Worsted Cloth Manufacture.** By Roberts Beaumont, Assistant Lecturer at Yorkshire College, Leeds. 7s. 6d.

Others in preparation.

HISTORY, TOPOGRAPHY, &c.

Rome and the Campagna. By R. Burn, M.A. With 85 Engravings and 26 Maps and Plans. With Appendix. 4to. 3l. 3s.

Old Rome. A Handbook for Travellers. By R. Burn, M.A. With Maps and Plans. Demy 8vo. 10s. 6d.

Modern Europe. By Dr. T. H. Dyer. 2nd Edition, revised and continued. 5 vols. Demy 8vo. 2l. 12s. 6d.

The History of the Kings of Rome. By Dr. T. H. Dyer. 8vo. 16s.

The History of Pompeii: its Buildings and Antiquities. By T. H. Dyer. 3rd Edition, brought down to 1874. Post 8vo. 7s. 6d.

The City of Rome: its History and Monuments. 2nd Edition, revised by T. H. Dyer. 5s.

Ancient Athens: its History, Topography, and Remains. By T. H. Dyer. Super-royal 8vo. Cloth. 1l. 5s.

The Decline of the Roman Republic. By G. Long. 5 vols. 8vo. 14s. each.

A History of England during the Early and Middle Ages. By C. H. Pearson, M.A. 2nd Edition revised and enlarged. 8vo. Vol. I. 16s. Vol. II. 14s.

Historical Maps of England. By C. H. Pearson. Folio. 3rd Edition revised. 31s. 6d.

History of England, 1800–46. By Harriet Martineau, with new and copious Index. 5 vols. 3s. 6d. each.

A Practical Synopsis of English History. By A. Bowes. 9th Edition, revised. 8vo. 1s.

Lives of the Queens of England. By A. Strickland. Library Edition, 8 vols. 7s. 6d. each. Cheaper Edition, 6 vols. 5s. each. Abridged Edition, 1 vol. 6s. 6d. Mary Queen of Scots, 2 vols. 5s. each. Tudor and Stuart Princesses, 5s.

Eginhard's Life of Karl the Great (Charlemagne). Translated, with Notes, by W. Glaister, M.A., B.C.L. Crown 8vo. 4s. 6d.

The Elements of General History. By Prof. Tytler. New Edition, brought down to 1874. Small Post 8vo. 3s. 6d.

History and Geography Examination Papers. Compiled by C. H. Spence, M.A., Clifton College. Crown 8vo. 2s. 6d.

PHILOLOGY.

WEBSTER'S DICTIONARY OF THE ENGLISH LANGUAGE. With Dr. Mahn's Etymology. 1 vol. 1628 pages, 3000 Illustrations. 21s.; half calf, 30s.; calf or half russia, 31s. 6d.; russia, 2l. With Appendices and 70 additional pages of Illustrations, 1919 pages, 31s. 6d.; half calf, 2l.; calf or half russia, 2l. 2s.; russia, 2l. 10s.

'THE BEST PRACTICAL ENGLISH DICTIONARY EXTANT.'—*Quarterly Review*, 1873.

Prospectuses, with specimen pages, post free on application.

Richardson's Philological Dictionary of the English Language. Combining Explanation with Etymology, and copiously illustrated by Quotations from the best Authorities. With a Supplement. 2 vols. 4to. 4l. 14s. 6d. Supplement separately. 4to. 12s.

Brief History of the English Language. By Prof. James Hadley, LL.D., Yale College. Fcap. 8vo. 1s.

The Elements of the English Language. By E. Adams, Ph.D. 21st Edition. Post 8vo. 4s. 6d.

Philological Essays. By T. H. Key, M.A., F.R.S. 8vo. 10s. 6d.

Language, its Origin and Development. By T. H. Key, M.A., F.R.S. 8vo. 14s.

Synonyms and Antonyms of the English Language. By Archdeacon Smith. 2nd Edition. Post 8vo. 5s.

Synonyms Discriminated. By Archdeacon Smith. Demy 8vo. 2nd Edition revised. 14s.

Bible English. Chapters on Words and Phrases in the Bible and Prayer Book. By Rev. T. L. O. Davies. 5s.

The Queen's English. A Manual of Idiom and Usage. By the late Dean Alford 6th Edition. Fcap. 8vo. 1s. sewed. 1s. 6d. cloth.

A History of English Rhythms. By Edwin Guest, M.A., D.C.L. LL.D. New Edition, by Professor W. W. Skeat. Demy 8vo. 18s.

Elements of Comparative Grammar and Philology. For Use in Schools. By A. C. Price, M.A., Assistant Master at Leeds Grammar School. Crown 8vo. 2s. 6d.

Questions for Examination in English Literature. By Prof. W. W. Skeat. 2nd Edition, revised. 2s. 6d.

Etymological Glossary of nearly 2500 English Words derived from the Greek. By the Rev. E. J. Boyce. Fcap. 8vo. 3s. 6d.

A Syriac Grammar. By G. Phillips, D.D. 3rd Edition, enlarged. 8vo. 7s. 6d.

DIVINITY, MORAL PHILOSOPHY, &c.

BY THE REV. F. H. SCRIVENER, A.M., LL.D., D.C.L.

Novum Testamentum Græce. Editio major. Being an enlarged Edition, containing the Readings of Westcott and Hort, and those adopted by the Revisers, &c. 7*s.* 6*d.* *For other Editions see page 3.*

A Plain Introduction to the Criticism of the New Testament. With Forty Facsimiles from Ancient Manuscripts. 3rd Edition. 8vo. 18*s.*

Six Lectures on the Text of the New Testament. For English Readers. Crown 8vo. 6*s.*

Codex Bezæ Cantabrigiensis. 4to. 26*s.*

The New Testament for English Readers. By the late H. Alford, D.D. Vol. I. Part I. 3rd Edit. 12*s.* Vol. I. Part II. 2nd Edit. 10*s.* 6*d.* Vol. II. Part I. 2nd Edit. 16*s.* Vol. II. Part II. 2nd Edit. 16*s.*

The Greek Testament. By the late H. Alford, D.D. Vol. I. 7th Edit. 1*l.* 8*s.* Vol. II. 8th Edit. 1*l.* 4*s.* Vol. III. 10th Edit. 18*s.* Vol. IV. Part I. 5th Edit. 18*s.* Vol. IV. Part II. 10th Edit. 14*s.* Vol. IV. 1*l.* 12*s.*

Companion to the Greek Testament. By A. C. Barrett, M.A. 5th Edition, revised. Fcap. 8vo. 5*s.*

The Book of Psalms. A New Translation, with Introductions, &c. By the Very Rev. J. J. Stewart Perowne, D.D. 8vo. Vol. I. 6th Edition, 18*s.* Vol. II. 6th Edit. 16*s.*

——— Abridged for Schools. 6th Edition. Crown 8vo. 10*s.* 6*d.*

History of the Articles of Religion. By C. H. Hardwick. 3rd Edition. Post 8vo. 5*s.*

History of the Creeds. By J. R. Lumby, DD. 3rd Edition. Crown 8vo. 7*s.* 6*d.*

Pearson on the Creed. Carefully printed from an early edition. With Analysis and Index by E. Walford, M.A. Post 8vo. 5*s.*

Liturgies and Offices of the Church, for the Use of English Readers, in Illustration of the Book of Common Prayer. By the Rev. Edward Burbidge, M.A. Crown 8vo. 9*s.*

An Historical and Explanatory Treatise on the Book of Common Prayer By Rev. W. G. Humphry, B.D. 6th Edition, enlarged. Small Post 8vo. 2*s.* 6*d.*; Cheap Edition, 1*s.*

A Commentary on the Gospels, Epistles, and Acts of the Apostles. By Rev. W. Denton, A.M. New Edition. 7 vols. 8vo. 9*s.* each.

Notes on the Catechism. By Rt. Rev. Bishop Barry. 8th Edit. Fcap. 2*s.*

The Winton Church Catechist. Questions and Answers on the Teaching of the Church Catechism. By the late Rev. J. S. B. Monsell, LL.D. 4th Edition. Cloth, 3*s.*; or in Four Parts, sewed.

The Church Teacher's Manual of Christian Instruction. By Rev. M. F. Sadler. 38th Thousand. 2*s.* 6*d.*

FOREIGN CLASSICS.

A Series for use in Schools, with English Notes, grammatical and explanatory, and renderings of difficult idiomatic expressions. Fcap. 8vo.

Schiller's Wallenstein. By Dr. A. Buchheim. 5th Edit. 5*s.* Or the Lager and Piccolomini, 2*s.* 6*d.* Wallenstein's Tod, 2*s.* 6*d.*

——— **Maid of Orleans.** By Dr. W. Wagner. 2nd Edit. 1*s.* 6*d.*

——— **Maria Stuart.** By V. Kastner. 2nd Edition. 1*s.* 6*d.*

Goethe's Hermann and Dorothea. By E. Bell, M.A., and E. Wölfel. 1*s.* 6*d.*

German Ballads, from Uhland, Goethe, and Schiller. By C. L. Bielefeld. 3rd Edition. 1*s.* 6*d.*

Charles XII., par Voltaire. By L. Direy. 7th Edition. 1*s.* 6*d.*

Aventures de Télémaque, par Fénélon. By C. J. Delille. 4th Edition. 2*s.* 6*d.*

Select Fables of La Fontaine. By F. E. A. Gasc. 18th Edit. 1*s.* 6*d.*

Picciola, by X. B. Saintine. By Dr. Dubuc. 15th Thousand. 1*s.* 6*d.*

Lamartine's Le Tailleur de Pierres de Saint-Point. By J. Boïelle, 4th Thousand. Fcap. 8vo. 1*s.* 6*d.*

Italian Primer. By Rev. A. C. Clapin, M.A. Fcap. 8vo. 1*s.*

FRENCH CLASS-BOOKS.

French Grammar for Public Schools. By Rev. A. C. Clapin, M.A. Fcap. 8vo. 11th Edition, revised. 2*s.* 6*d.*

French Primer. By Rev. A. C. Clapin, M.A. Fcap. 8vo. 7th Ed. 1*s.*

Primer of French Philology. By Rev. A. C. Clapin. Fcap. 8vo. 3rd Edit. 1*s.*

Le Nouveau Trésor; or, French Student's Companion. By M. E. S. 18th Edition. Fcap. 8vo. 1*s.* 6*d.*

French Examination Papers in Miscellaneous Grammar and Idioms. Compiled by A. M. M. Stedman, M.A. 2nd Edition, revised. Crown 8vo. 2*s.* 6*d.*

Key to the above. By G. A. Schrumpf, Univ. of France. Crown 8vo. 5*s.* (For Teachers or Private Students only.)

Manual of French Prosody. By Arthur Gosset, M.A. Crown 8vo. 3*s.*

Wortfolge, or Rules and Exercises on the Order of Words in German Sentences. By Dr. F. Stock. 1s. 6d.

A German Grammar for Public Schools. By the Rev. A. C. Clapin and F. Holl Müller. 4th Edition. Fcap. 2s. 6d.

A German Primer, with Exercises. By Rev. A. C. Clapin. 1s.

Kotzebue's Der Gefangene. With Notes by Dr. W. Stromberg. 1s.

German Examination Papers in Grammar and Idiom. By R. J. Morich. 2s. 6d. [*Key in the press.*

MODERN GERMAN SCHOOL CLASSICS.

Small Crown 8vo.

German Examination Course. By Prof. F. Lange, Ph.D. Elementary, 2s. Intermediate, 2s. Advanced, 1s. 6d.

Hey's Fabeln Für Kinder. Edited by Prof. F. Lange, Ph.D. 1s. 6d.

Bechstein's Märchen. Edited by Prof. H. Hager, Ph.D.

Benedix's Dr. Wespe. Edited by F. Lange, Ph.D. 2s. 6d.

Schiller's Jugendjahre. Edited by Prof. H. Hager, Ph.D.

Hoffman's Meister Martin, der Küfner. By Prof. F. Lange, Ph.D. 1s. 6d.

Heyse's Hans Lange. By A. A. Macdonell, M.A., Ph.D. 2s.

Auerbach's Auf Wache, and Roquette's Der Gefrorene Kuss. By A. A. Macdonell, M.A. 2s.

Moser's Der Bibliothekar. By Prof. F. Lange, Ph.D. 2s.

Eber's Eine Frage. By F. Storr, B.A. 2s.

Freytag's Die Journalisten. By Prof. F. Lange, Ph.D. 2s. 6d.

Gutzkow's Zopf und Schwert. By Prof. F. Lange, Ph.D. 2s.

German Epic Tales. Edited by Karl Neuhaus, Ph.D. 2s. 6a.

Humoresken. Novelletten der besten deutschen Humoristen der Gegenwart. Edited by A. A. Macdonell, M.A. Oxon. Authorised Edition. [*In preparation.*

ENGLISH CLASS-BOOKS.

Comparative Grammar and Philology. By A. C. Price, M.A., Assistant Master at Leeds Grammar School. 2s. 6d.

The Elements of the English Language. By E. Adams, Ph.D. 21st Edition. Post 8vo. 4s. 6d.

The Rudiments of English Grammar and Analysis. By E. Adams, Ph.D. 16th Thousand. Fcap. 8vo. 1s.

A Concise System of Parsing. By L. E. Adams, B.A. 1s. 6d.

Examples for Grammatical Analysis (Verse and Prose). Selected, &c., by F. Edwards. 6d.

Notes on Shakespeare's Midsummer Night's Dream. By T. Duff Barnett, B.A. 1s. Julius Cæsar, 1s. Henry V., 1s.

By C. P. Mason, Fellow of Univ. Coll. London.

First Notions of Grammar for Young Learners. Fcap. 8vo. 35th to 40th Thousand. Cloth. 9d.

First Steps in English Grammar for Junior Classes. Demy 18mo. 44th Thousand. 1s.

Outlines of English Grammar for the Use of Junior Classes. 65th to 70th Thousand. Crown 8vo. 2s.

English Grammar, including the Principles of Grammatical Analysis. 30th Edition. 120th to 124th Thousand. Crown 8vo. 3s. 6d.

A Shorter English Grammar, with copious Exercises. 30th to 33rd Thousand. Crown 8vo. 3s. 6d.

English Grammar Practice, being the Exercises separately. 1*s*.

Code Standard Grammars. Parts I. and II., 2*d*. each. Parts III., IV., and V., 3*d*. each.

Notes of Lessons, their Preparation, &c. By José Rickard, Park Lane Board School, Leeds, and A. H. Taylor, Rodley Board School, Leeds. 2nd Edition. Crown 8vo. 2s. 6*d*.

A Syllabic System of Teaching to Read, combining the advantages of the 'Phonic' and the 'Look-and-Say' Systems. Crown 8vo. 1s.

Practical Hints on Teaching. By Rev. J. Menet, M.A. 6th Edit. revised. Crown 8vo. paper, 2s.

How to Earn the Merit Grant. A Manual of School Management. By H. Major, B.A., B.Sc. 2nd Edit. revised. Part I. Infant School, 3s. Part II. 4s. Complete, 6s.

Test Lessons in Dictation. 4th Edition. Paper cover, 1*s*. 6*d*.

Drawing Copies. By P. H. Delamotte. Oblong 8vo. 12*s*. Sold also in parts at 1s. each.

Poetry for the Schoolroom. New Edition. Fcap. 8vo. 1*s*. 6*d*.

The Botanist's Pocket-Book. With a copious Index. By W. R. Hayward. 5th Edition, revised. Crown 8vo. cloth limp. 4s. 6d.

Experimental Chemistry, founded on the Work of Dr. Stöckhardt. By C. W. Heaton. Post 8vo. 5s.

Lectures on Musical Analysis. Sonata-form, Fugue. Illustrated from Classical Masters. By Prof. H. C. Banister. 7*s*. 6*d*.

GEOGRAPHICAL SERIES. By M. J. Barrington Ward, M.A.

The Map and the Compass. A Reading-Book of Geography. For Standard I. 6d.

The Round World. A Reading-Book of Geography. For Standard II. 10*d*.

The Child's Geography. For the Use of Schools and for Home Tuition. 6*d*.

The Child's Geography of England. With Introductory Exercises on the British Isles and Empire, with Questions. 2s. 6*d*. Without Questions, 2s.

Geography Examination Papers. (See History and Geography Papers, p. 12.)

Helps' Course of Poetry, for Schools. A New Selection from the English Poets, carefully compiled and adapted to the several standards by E. A. Helps, one of H.M. Inspectors of Schools.

Book I. Infants and Standards I. and II. 134 pp. small 8vo. 9*d.*
Book II. Standards III. and IV. 224 pp. crown 8vo. 1*s.* 6*d.*
Book III. Standards V., VI., and VII. 352 pp. post 8vo. 2*s.*
Or in PARTS. Infants, 2*d.*; Standard I., 2*d.*; Standard II., 2*d.* Standard III., 4*d.*

Picture School-Books. In Simple Language, with numerous Illustrations. Royal 16mo.

The Infant's Primer. 3*d.*—School Primer. 6*d.*—School Reader. By J. Tilleard. 1*s.*—Poetry Book for Schools. 1*s.*—The Life of Joseph. 1*s.*—The Scripture Parables. By the Rev. J. E. Clarke. 1*s.*—The Scripture Miracles. By the Rev. J. E. Clarke. 1*s.*—The New Testament History. By the Rev. J. G. Wood, M.A. 1*s.*—The Old Testament History. By the Rev. J. G. Wood, M.A. 1*s.*—The Story of Bunyan's Pilgrim's Progress. 1*s.*—The Life of Martin Luther. By Sarah Crompton. 1*s.*

BOOKS FOR YOUNG READERS.

A Series of Reading Books designed to facilitate the acquisition of the power of Reading by very young Children. In 11 *vols. limp cloth,* 6*d. each.*

Those with an asterisk have a Frontispiece or other Illustration.

Suitable for Infants:

*The Two Parrots. A Tale of the Jubilee. By M. E. Wintle. 9 Illustrations.

*The Old Boathouse. Bell and Fan; or, A Cold Dip.

*Tot and the Cat. A Bit of Cake. The Jay. The Black Hen's Nest. Tom and Ned. Mrs. Bee.

*The Cat and the Hen. Sam and his Dog Redleg. Bob and Tom Lee. A Wreck.

*The New-born Lamb. The Rosewood Box. Poor Fan. Sheep Dog.

Suitable for Standards I. & II.:

*The Story of Three Monkeys.

*Story of a Cat. Told by Herself.

The Blind Boy. The Mute Girl. A New Tale of Babes in a Wood.

The Dey and the Knight. The New Bank Note. The Royal Visit. A King's Walk on a Winter's Day.

*Queen Bee and Busy Bee.

*Gull's Crag.

*A First Book of Geography. By the Rev. C. A. Johns. Illustrated. Double size, 1*s.*

Syllabic Spelling. By C. Barton. In Two Parts. Infants, 3*d.* Standard I., 3*d.*

Printed in Dunstable, United Kingdom

91220036R00167